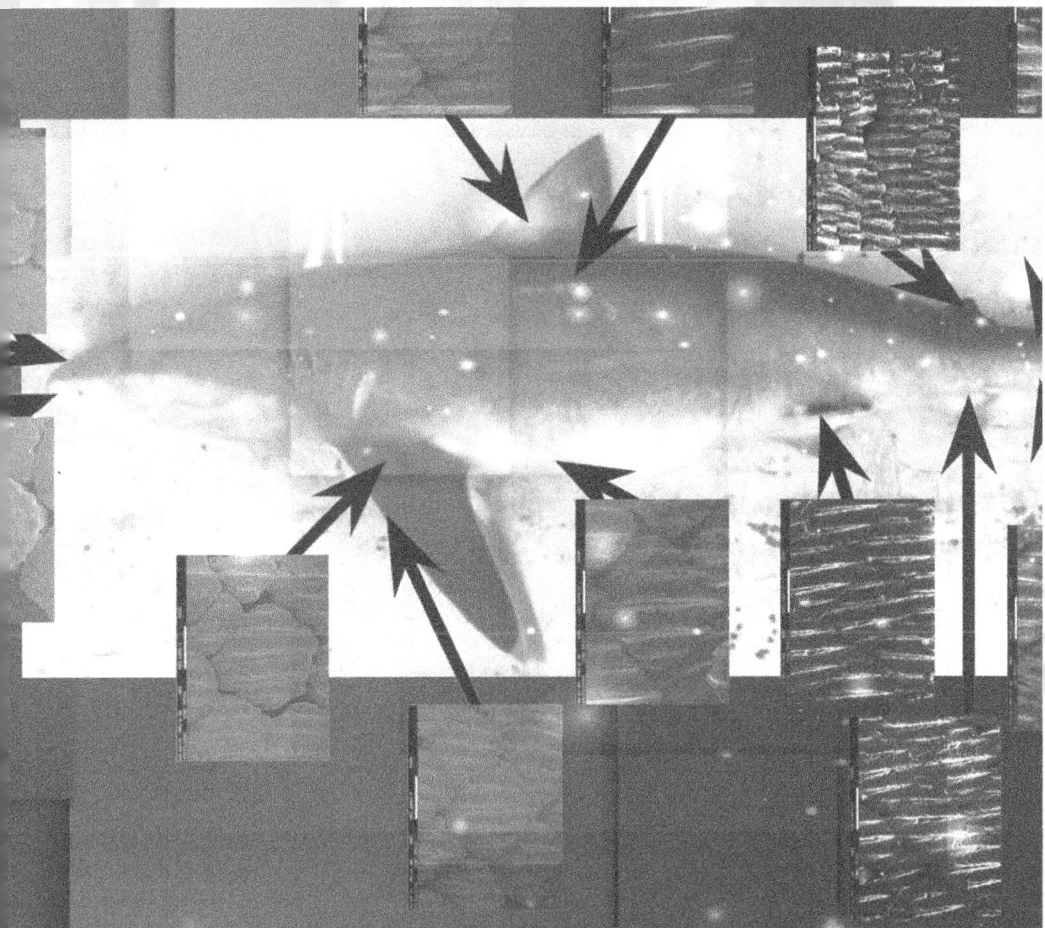

Industrial Biomimetics

Industrial Biomimetics

edited by

Akihiro Miyauchi
Masatsugu Shimomura

JENNY STANFORD

Published by

Jenny Stanford Publishing Pte. Ltd.
Level 34, Centennial Tower
3 Temasek Avenue
Singapore 039190

Email: editorial@jennystanford.com
Web: www.jennystanford.com

British Library Cataloguing-in-Publication Data
A catalogue record for this book is available from the British Library.

Industrial Biomimetics

ISBN 978-981-4800-07-5 (Hardcover)
ISBN 978-0-429-05883-7 (eBook)

Contents

Preface

Biomimetics is the innovative paradigm shift based on biodiversity for sustainability. Biodiversity is not only the result of evolutionary adaption but also the optimized solution of "an epic combinatorial chemistry" for sustainability, because the diversity has been acquired by "biological processes and technology," including "production processes," "operating principles," and "control systems," all of which differ from "human technology."

In this book, cutting-edge researchers have contributed related research results from the viewpoint of industrial applications of biomimetics. The chapters are categorized as follows: Chapters 1 to 8 are mechanical, Chapters 9 to 11 are optics, Chapter 12 is chemistry, and Chapters 13 to 15 are measurement. Readers can select chapters as per their areas of interest.

There are two types of applications of biomimetics. One starts with engineering tasks to solve an engineering problem using biomimetics, and the other starts with the knowledge of biology, applying it to engineering fields. This book discusses both approaches. We hope that this book will be an opportunity to expand the readers' ideas.

We thank the contributors of each chapter here and also thank Mr. Stanford Chong and Ms. Archana Ziradkar of Pan Stanford Publishing for promoting the publication of this book.

Akihiro Miyauchi

Masatsugu Shimomura

2019

Chapter 1

3D Modeling of Shark Skin and Prototype Diffuser for Fluid Control

Akihiro Miyauchi
Tokyo Medical and Dental University, 2-3-10 Kanda-Surugadai, Chiyodaku, Tokyo 101-0062, Japan
miyauchi.ibb@tmd.ac.jp

Sharks are well known for their fluid control ability, so shark skin has been widely researched [1–8]. Detailed 3D structure data can now be obtained due to the progress of advanced measurement methods such as X-ray micro–computed tomography (micro-CT). Computational fluid dynamics (CFD) has also been enhanced by the rapid progress of computational power. In this chapter, shark skin is taken as an example of biomimetic design (BMD) for industrial applications.

1.1 Analysis of Shark Skin

As pointed out by Vincent et al., the ways of building functional structures are different between biology and technology [9]. The surfaces of living matter have evolved into various functional 3D

Industrial Biomimetics
Edited by Akihiro Miyauchi and Masatsugu Shimomura

ISBN 978-981-4800-07-5 (Hardcover), 978-0-429-05883-7 (eBook)
www.jennystanford.com

structures that are beyond human approach [6]. One example is the placoid scales (dermal denticles) that cover the surface of shark skin, as shown in Fig. 1.1. As you can see, the denticles have groove structures on each scale. A shark mainly moves about one-third of its body (i.e., its tail) to swim. The tail scales at the tail have deeper grooves than those at the head, so the denticles seem to be structured for fluid control. Therefore, 3D measurement of the denticle shape is needed to reveal the mechanism of fluid control.

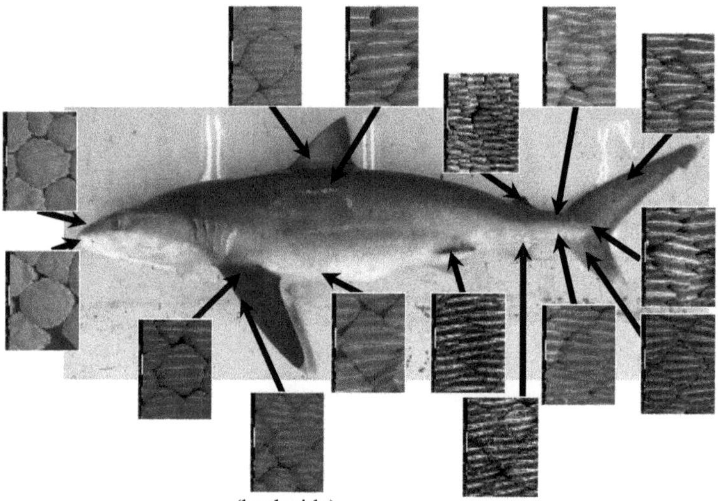

Figure 1.1 Scanning electron microscopy (SEM) images of denticles on the silvertip shark.

X-ray micro–computed tomography (micro-CT) is a powerful tool to measure the 3D features at the micron scale. X-ray micro-CT measures a lot of cross sections of a specimen in microscale resolution. The 3D topography of denticles can be obtained by image synthesis. Figure 1.2 shows the four kinds of shark skins and denticles measured by X-ray micro-CT. Single denticles are extracted from shark skin (aggregate of denticles) by image processing. Quantitative analysis of denticle features is made possible by the synthesis of CT images. The denticle sizes and groove shapes differ for each species. Each species has groove structures on the denticle surface, which is why shark skin is referred to as riblet (simple groove structure).

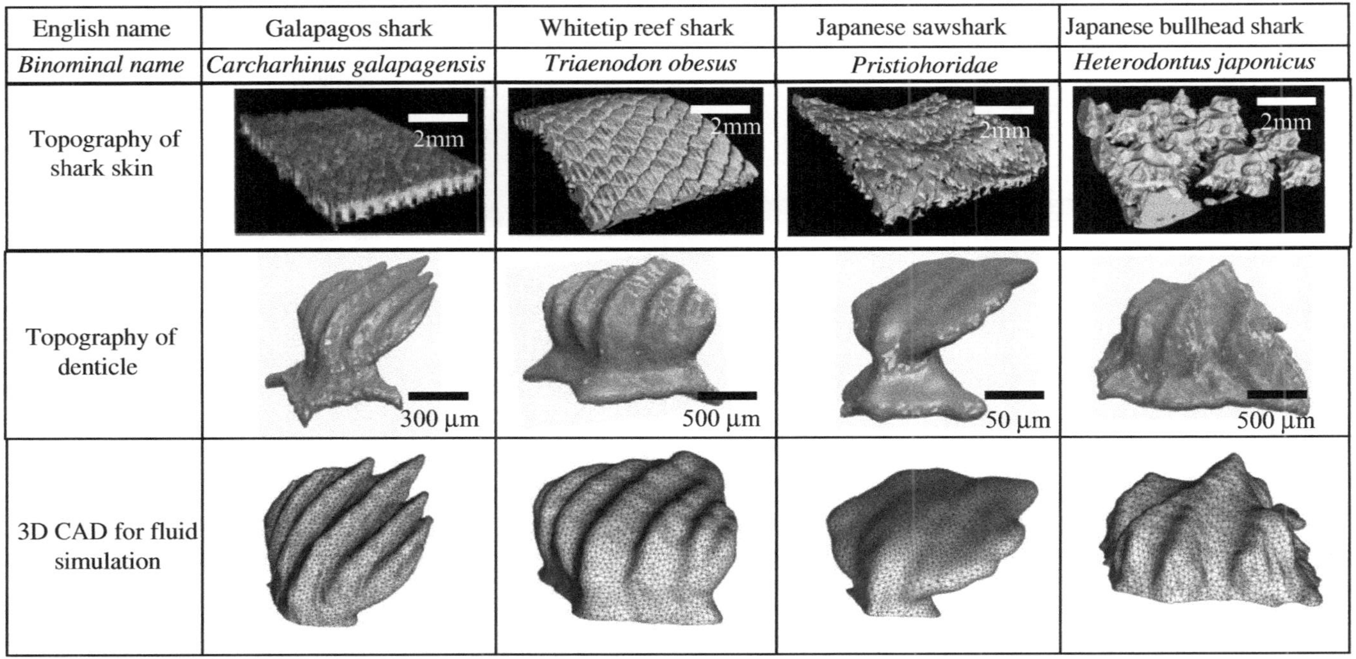

Figure 1.2 Tomography of shark skins and denticles measured by X-ray micro-CT, and 3D CAD of denticles.

1.2 Biomimetic Design of Shark Skin

The conventional models of shark skin are riblet structures that have been reported to reduce drag [1–8]. In this section, shark skin is modeled on the basis of precise morphological analysis by X-ray micro-CT and computational fluid dynamics (CFD).

The data of tomography obtained by X-ray micro-CT contains inner structure information of denticles, but only surface morphology is necessary for 3D fluid simulations. Thus, on the basis of the denticle tomography data, a 3D surface mesh of each denticle is made, as shown in the bottom column of Fig. 1.2. By using this 3D computer-aided design (CAD) data, fluid states near each denticle are analyzed.

Figure 1.3 shows the fluid model for denticle analysis. The eight denticles are located on the bottom of the model. The water flows along the *x* direction at 1 m/s. The flow states near the denticles are analyzed.

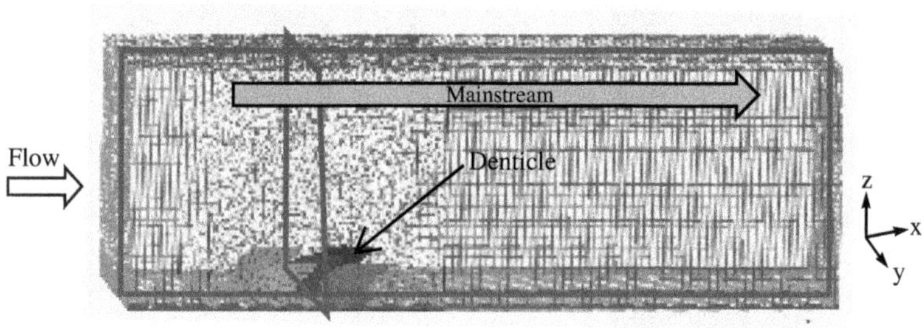

Figure 1.3 Flow analysis model for denticles.

Figure 1.4 shows the flow vectors in an *x-z* cross section of Galapagos shark denticles. The upstream vectors appear. This flow is caused by the tilted structure of denticles and is called rising flow, Uz. The same result was reported by Luo et al. [8]. The generation of Uz is obvious because the mainstream conflicts at the slope structure. CFD revealed that Uz increases with slope angle but a vortex is generated at the bottom of slope when the slope angle exceeds about 50°. The tilt angles of shark denticles shown in Fig. 1.2 are about 20°–30°. This seems to avoid the energy loss induced by vortex generation during the heaving motion of swimming. Therefore, the slope angle for biomimetic design (BMD) was fixed at 25°.

Figure 1.4 Rising flows by denticles in the *x-z* cross section (Galapagos shark).

Another feature of the flow state is the longitudinal vortex, wx, as shown in Fig. 1.5. Longitudinal vortexes appear in the *y-z* plane, and the rotation axis is along the *x* direction. Both rising flow and longitudinal vortex appear in three species in Fig. 1.2.

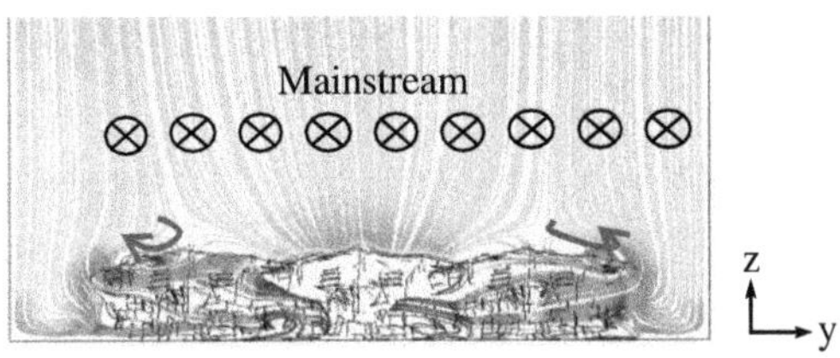

Figure 1.5 Longitudinal vortex near the grooves of denticle surface in the *y-z* cross section (Galapagos shark).

The generation model of the longitudinal vortex is shown in Fig. 1.6. Imagine that the main flow paths are through the denticle groove. The flows contact the groove walls and receive the drag. The drag depends on the contact area, so a small wall surface (S_1) causes less drag than a large wall surface (S_2). The flow velocity $\mathbf{v}_1$ near the surface S_1 becomes faster than $\mathbf{v}_2$. The product of pressure p and velocity $\mathbf{v}$ is constant, so the pressure p_1 becomes less than p_2. This pressure difference in the *y-z* plane causes the longitudinal vortex. To verify this vortex generation model, the flows near the asymmetric groove structures were investigated.

Figure 1.7a shows the investigated fluid models. The fluid configuration is the rectangular case. The asymmetric grooves were formed on the case bottom. Two types of asymmetric groove structures were investigated: asymmetry in (Fig. 1.7b) groove heights and (Fig. 1.7c) groove widths. Both types have different side wall areas: S_1 and S_2 shown in Fig. 1.6.

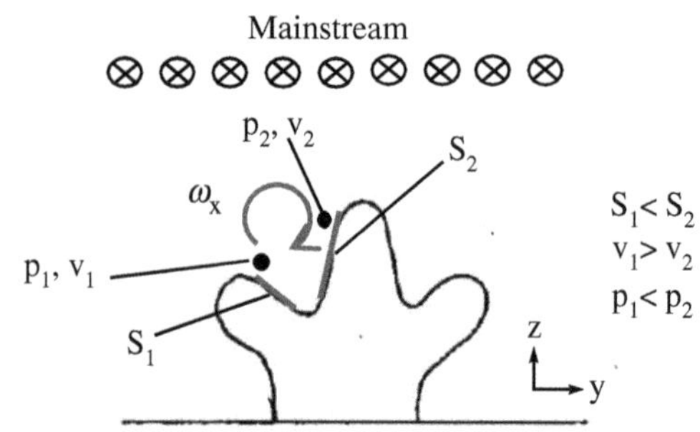

Figure 1.6 Generation model of a longitudinal vortex.

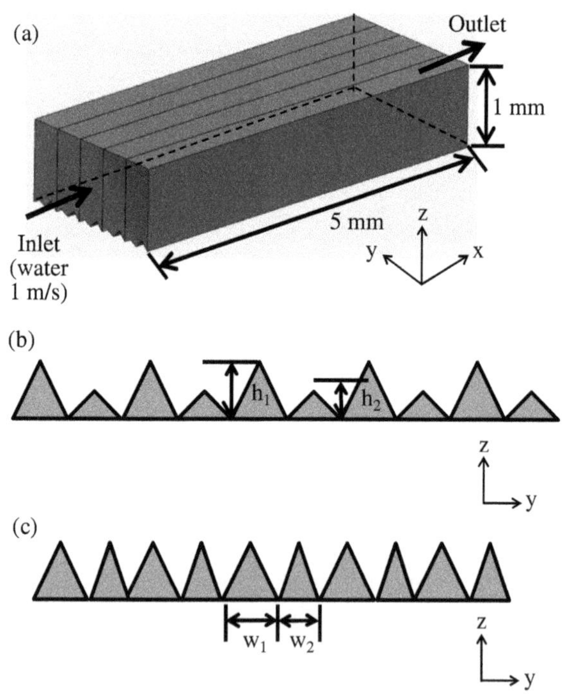

Figure 1.7 Investigated model of an asymmetric groove. (a) Fluid model, (b) case of asymmetric groove heights (w_1, w_2, h_1 = 0.1 mm), and (c) case of asymmetric groove widths (h_1, h_2, w_1 = 0.1 mm).

Figure 1.8 shows the simulation results for the longitudinal vortex. The increase of w_x from symmetric grooves is shown as a function of h_1/h_2 and w_1/w_2, where h_i and w_i are groove heights and widths, respectively. The difference of w_x between the symmetric and the asymmetric groove, $\delta\omega_x$, increases with h_1/h_2 and w_1/w_2. Comparing Fig. 1.8a and Fig. 1.8b, the asymmetry of the groove height is more effective than that of the width to enhance ω_x. The h_1/h_2 and w_1/w_2 of the Galapagos shark denticle are around 2.0 and 1.5, respectively. Therefore, the denticles seem to evolve to generate ω_x effectively.

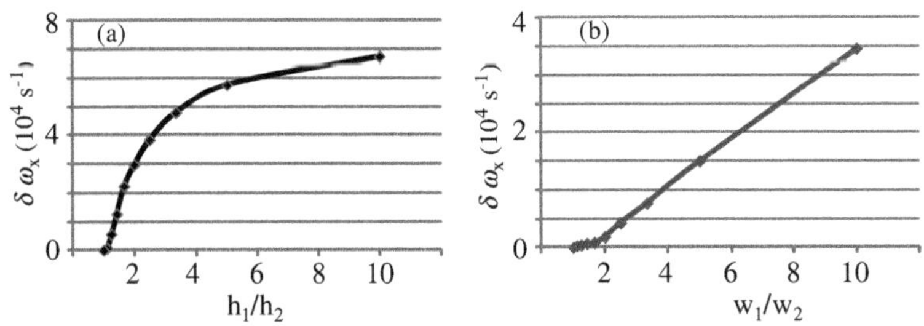

Figure 1.8 Simulation results of a longitudinal vortex, w_x, as functions of groove height and width ratio. Both cases indicate that asymmetry of the groove structure enhance the generation of a longitudinal vortex. 1.0 of horizontal axes indicate the symmetric structures.

On the basis of the flow analysis on slope and asymmetric groove structures, three kinds of surface structure are designed, as shown in Fig. 1.9. All slopes are 25°. The grooves in Fig. 1.9b and Fig. 1.9c have asymmetry for heights and widths, respectively. The fluid simulations showed that these structures generated rising flows, Uz, and longitudinal vortexes, wx.

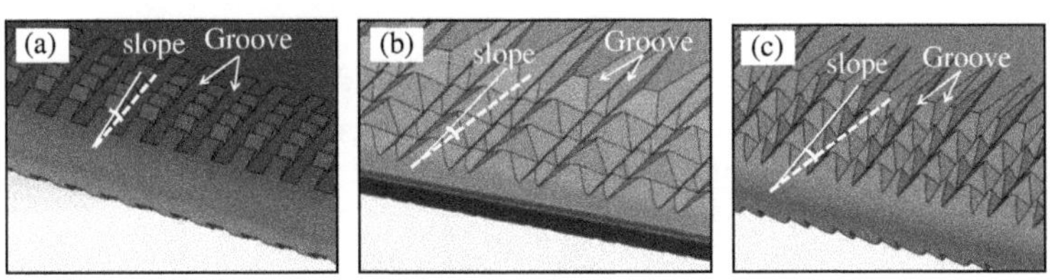

Figure 1.9 Designed surface structures based on CFD analysis of shark skin. Features are slope and asymmetric grooves.

In this section, the flow states near the denticles are analyzed. A 3D morphology was obtained by X-ray micro-CT and revealed rising flows and longitudinal vortexes near the denticles. These flow features are generated by a slope and asymmetric groove structures. This means that the similar flow states can be obtained by designing a slope and asymmetric structure. This surface design method (namely observing living creatures, understanding the mechanism of their functionality, and designing a structure that can reproduce the same function) is named "biomimetic design."

1.3 Evaluation of Biomimetic Design

The centrifugal compressor shown in Fig. 1.10 is an important mechanical component in industrial processes such as petroleum and chemical processing and iron manufacturing and energy infrastructure such as gas turbines and natural gas pipelines [10]. The suction power is generated by the impeller and diffuser. The impeller is rotated with a rotor at several ten thousand revolutions per minute (rpm). The rotating impeller makes the swirl flow. The swirl flow is connected to the diffuser, and the kinetic energy of the swirl gas flow is partially converted to the gas pressure. The diffuser configuration is shown in Fig. 1.11. This diffuser has vanes, so it is called a vaned diffuser. The gas velocity reaches Mach 0.6–0.8 at the front of the diffuser vanes. Therefore, instability of gas flow is often caused at the tips of vanes. Thus, BMD is used for vanes, and the diffuser characteristics are evaluated.

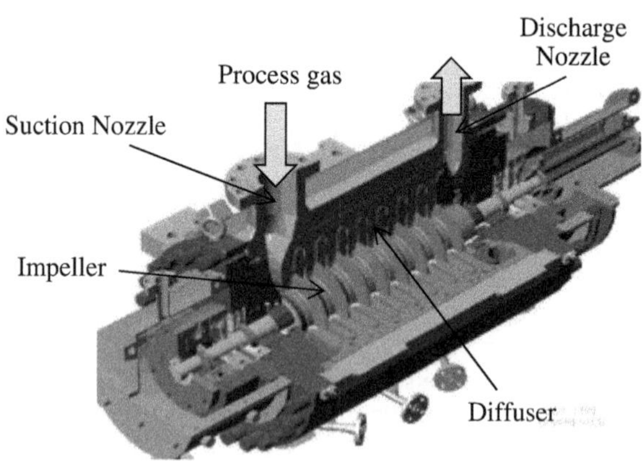

Figure 1.10 Inner structure of a centrifugal compressor.

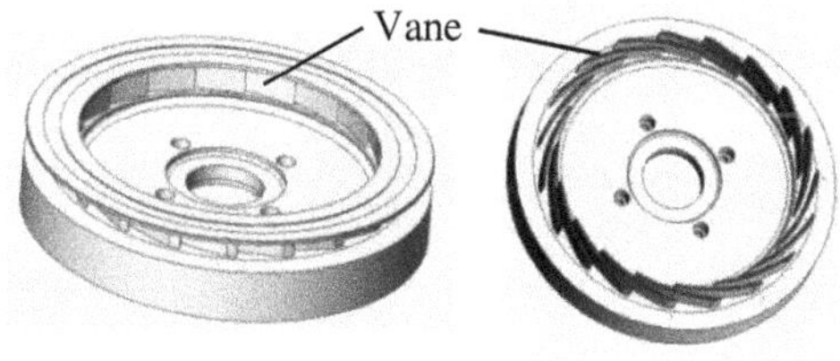

Figure 1.11 Appearance of a vaned diffuser.

Figure 1.12 shows the CAD and scanning electron microscopy (SEM) images of BMD. BMDs were fabricated on the tips of aluminum vanes by a precision cutting method. The original vane is shown in Fig. 1.12 as a reference. Comparing CADs and SEM images reveals that the fine patterns are fabricated on the aluminum tips. The sizes of fabricated patterns were measured by SEM. The errors between CAD and fabricated patterns in plane and depth were less than 10% and about 20%, respectively.

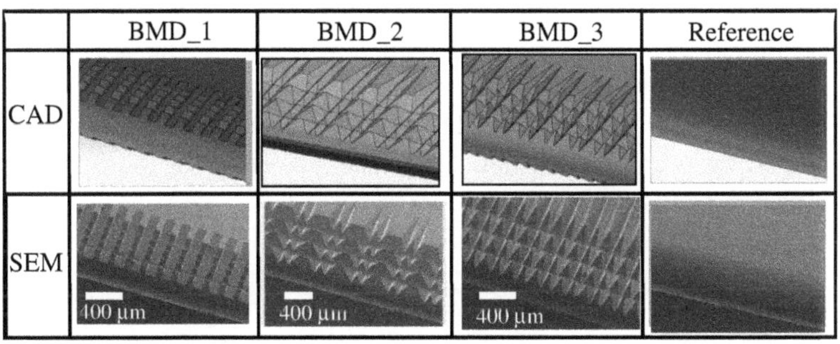

Figure 1.12 CAD and fabricated BMD structures on the front edge of vanes.

The BMDs were found to be able to fabricate on the tips of vanes, as shown in Fig. 1.12. However, a concavo-convex structure generally causes the pressure drag in fluid parts, so the surface structures were designed for not only the tip area but also the rest of the vane. Figure 1.13 shows the designed vane and SEM images of the fabricated surfaces. The BMD structure is located on the tip of the vane, and the riblet structure covers the rest of the whole surface. The riblet structure reduces the pressure drag [1, 3, 4], so it compensates for the pressure drag induced by the BMD structure.

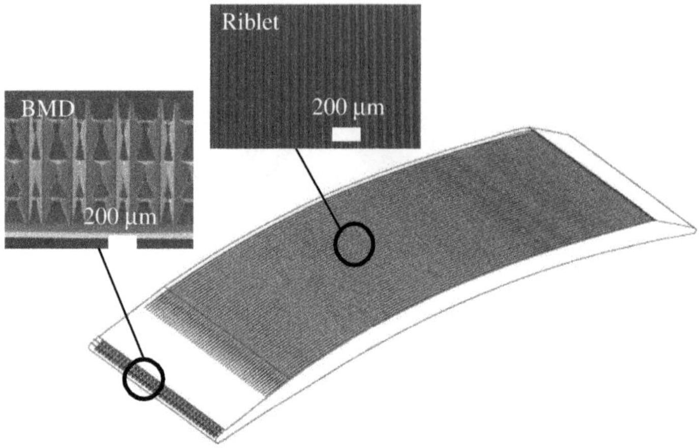

Figure 1.13 Designed vane. BMD area and riblet structure are on the vane surface.

The BMD and riblet structures were confirmed to be successfully fabricated, so the diffusers with these fabricated vanes were assembled as a BMD diffuser. Figure 1.14 shows the assembled diffuser. The prototype diffuser has 19 vanes made of aluminum. Its body is made of acrylonitrile-butadiene-styrene (ABS) resin and shaped by precision cutting. The vanes are stuck to the body by adhesion bonds.

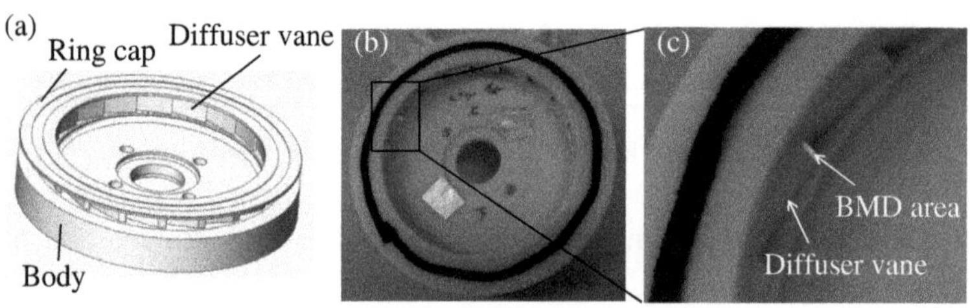

Figure 1.14 Assembled BMD vaned diffuser. (a) Designed CAD of a diffuser, (b) appearance of an assembled prototype diffuser, and (c) magnified image of BMD-designed vanes.

The fluid characteristics of the assembled diffusers were evaluated by using compressor evaluation equipment. The evaluation equipment can generate a high-speed gas flow by rotating the impeller. The gas is introduced to the assembled diffuser. The flow volume is controlled by the flow control valve located at the gas inlet side.

Figure 1.15 shows the pressure recovery characteristics of the BMD diffuser and conventional diffuser. The conventional diffuser has plane vanes stalled at a low flow rate, that is, in the high-gas-velocity region. However, the diffuser with BMD and riblet has a wider stable range, that is, a large surge margin. This feature is attractive for improving compressor ability because it enables the compressor to be controlled in a wider range of plant operation conditions.

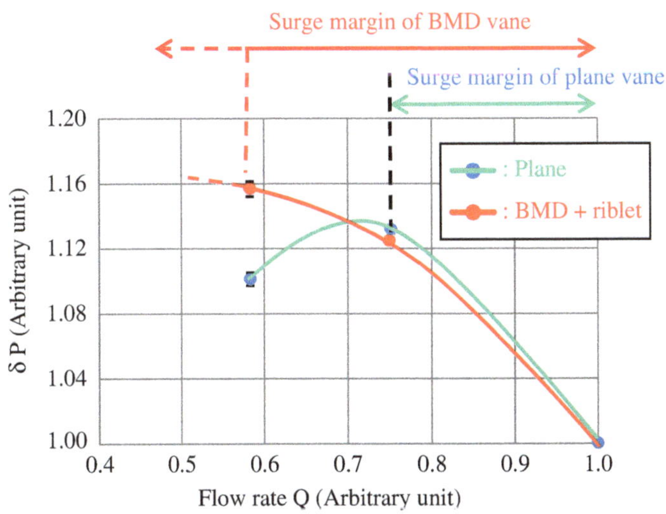

Figure 1.15 Pressure recovery characteristics of diffuser. A BMD vane diffuser indicates a wider range of the surge margin than a conventional vane diffuser.

1.4 Summary

Shark skin structures were analyzed in the micron scale by using X-ray micro-CT. Functions of shark skin in fluid were analyzed using a 3D denticle model, and two kinds of featured flow states were found: rising flow and longitudinal vortex. An artificial surface structure was investigated that can generate the same flow states as shark skin and be manufactured industrially. This design sequence (analyze a living thing's function and reproduce it artificially) is named BMD. BMDs of shark skin were proposed, and prototype diffusers were made by using a precise cutting method. The improvement of process conditions was shown by measurements of BMD-embedded diffusers. The BMDs of shark skin will be expanded in not only fluid dynamics but also biological fields [11].

Acknowledgments

The author thanks Drs. Moriya, Miyazaki, Honda, and Hiradate of Hitachi Ltd., as well as Mr. Tokui of Tempstaff, for fluid simulation, experiments and discussions. The author also thanks Dr. Shinohara of the National Museum of Nature and Science of Japan, Drs. Tomita and Sato of Churaumi Aquarium for supplying shark skin specimens, and Prof. Shimomura and Dr. Hirai of Chitose Institute of Science and Technology for X-ray micro-CT measurements and fruitful discussions. A part of this article was supported by the New Energy Development Organization (NEDO) of Japan.

References

1. Choi, H., Moin, P. and Kim, J. (1993). Direct numerical simulation of turbulent flow over riblets, *J. Fluid Mech.*, **255,** pp. 503–539.
2. Bechert, D. W., Bruse, M., Hage, W. and Meyer, R. (2000). Fluid mechanics of biological surfaces and their technological application, *Naturwissenshaften*, **87**, pp. 157–171.
3. Lee, S.-J. and Lee, S.-H. (2001). Flow field analysis of a turbulent boundary layer over a riblet surface, *Exp. Fluids*, **30**, pp. 153–166.
4. Dean, B. and Bhushan, B. (2010). Shark-skin surfaces for fluid-drag reduction in turbulent flow: a review, *Phil. Trans. R. Soc.* **A 368**, pp. 4775–4806.
5. Buttner, C. C. and Schulz, U. (2011). Shark skin inspired riblet structures as aerodynamically optimized high temperature coatings for blades of aeroengines, *Smart Mater. Struct.*, **20**, 094016.
6. Bhushan, B. (2012). Bioinspired structured surfaces, *Langmuir*, **28**, pp. 1698–1714.
7. Wen, L., Weaver, J. C. and Lauder, G. V. (2014). Biomimetic shark skin: design, fabrication and hydrodynamic function, *J. Experimental Biology*, **217**, pp. 1656–1666.

8. Luo, Y., Liu, Y., Zhang, D. and Ng, E. Y. K. (2014). Influence of morphology for drag reduction effect of sharkskin surface, *J. Mech. Med. Biol.*, **14**(2), 1450029.
9. Vincent, J. F. V., Bogatyreva, O. A., Bogatyrev, N. R., Bowyer, A. and Pahl, A.-K. (2006). Biomimetics: its practice and theory, *J. R. Soc. Interface*, **3**, 471–482.
10. Sorokes, J. M. and Kuzdzal, M. J. (2010). Centrifugal compressor evolution, *Proc. 39th Turbomachinery Symp.*, pp. 59–70.
11. Chung, K. K., Schumacher, J. F., Sampson, E. M., Bume, R. A., Antonelli, P. J. and Brennan, A. B. (2007). Impact of engineered surface microtopography on biofilm formation of *Staphyloccus aureus*, *Biointerphases*, **2**(2), pp. 89–94.

Chapter 2

Friction Control Surfaces in Nature: Analysis of Firebrat Scales

Yuji Hirai, Naoto Okuda, and Masatsugu Shimomura
Department of Applied Chemistry and Bioscience, Chitose Institute of Science and Technology, 758-65, Bibi, Chitose, Hokkaido 066-8655, Japan
y-hirai@photon.chitose.ac.jp

2.1 Nature Friction Control Surface

Friction is an important issue with regard to saving energy and preventing the wear of parts in a wide variety of fields, such as the automobile industry [1] and medicine [2]. Nowadays, lubricants are mainly used for reducing friction forces; however, other technologies are now in demand as lubricants generally lead to pollution of the environment [3]. In recent years, surface micro- and nanostructures have attracted a good deal of attention due to their potential to reduce friction [4–6]. Although there are a lot of reports on friction reduction by micro- and nanostructures, most effective patterns remain under consideration as friction is a complex phenomenon affected by atomic-level structure [7, 8]. Meanwhile, friction is also important for living things, such as insect, animal, etc., so the surfaces

Industrial Biomimetics
Edited by Akihiro Miyauchi and Masatsugu Shimomura

ISBN 978-981-4800-07-5 (Hardcover), 978-0-429-05883-7 (eBook)
www.jennystanford.com

of living organisms would be adapted to reduce or protect their body surfaces during their evolution process. To research grazed surfaces of animals, some hints of superior designs may be found. In fact, some friction control surfaces of animals have been reported [9–12].

Terrestrial insects have developed their leg attachments into hairy pads and/or smooth pads [13, 14]. Both leg attachments match the surface profile of the habitat and then generate adhesive and friction force on the legs. Hairy attachments are specialized in adhesives, and there are a lot of reports described about their diversity and relation of body weight and density of hair structures [15]. And also, there is an interesting report about smooth pads (Fig. 2.1) [16]. Torrent frogs, bush crickets, etc., have smooth pads having hexagonally arranged microstructures, which act as a drainage ditch to remove water between leg surface and ground for preventing the hydroplaning effect. Prof. Gorb's group proved this function by preparing and measuring artificial pad structures. According to results, the microstructures allow stable movement of bush crickets on wet and slippery surfaces. Meanwhile, the pads have other functions working on dry surfaces. In dry conditions, usually stick–slip motion, repeatedly occurring static (high-) and kinetic (low-) friction states, happens when some objects drag. It's not so great for insects because they work hard to control their movements. However, when the leg surface has such surfaces, microstructures disperse static and kinetic friction to each microstructure, and it leads to constant friction force. As these strategies have been independently developed in the field of automotive industry, tire threads, it seems to be a natural advance.

In this section, the experiment that was conducted to find further, new friction control surface in nature is shown. This section focuses on the firebrat, *Thermobia domestica* [17–19], a small insect that lives in narrow spaces such as piled papers or bookshelves, because firebrat body surfaces are always in contact with other surfaces and are subject to wear. The firebrat is a type of primitive wingless insect belonging to the order Zygentoma, and its body surface is densely covered by scales with microgroove structures. Their surface scales might have evolved to reduce friction and to protect them from wear. Herein, surface observations and friction force measurements of the firebrat body surface by laser microscopy, field-emission scanning electron microscopy (FE-SEM), and colloidal probe atomic force microscopy (AFM) [20–22] are described. These results will contribute to the development of novel frictional surfaces applicable to a wide variety of industrial fields.

Figure 2.1 Photograph and SEM images of a bush cricket.

2.2 Firebrat Scale Surface Observations

Firebrats are easily bred and maintained in a laboratory. So fresh adults can be also collected for research. To use for investigation of the firebrat's scale properties, firebrats are kept in a refrigerator for 1 h in a state of suspended animation, sacrificed using ethyl acetate vapor, and then again kept in a refrigerator. Surface observations of firebrats are carried out by a laser microscope and FE-SEM after sputtering with Pt. Figure 2.2 shows the laser microscope image and FE-SEM images of the body surface of a firebrat (8.33 mm body length). The body surface of the firebrat is covered in dense scales, with the direction of growth of the scales at the head being different from those on other parts; scales on the head are oriented against the direction of forward movement, and growth directions of other parts are forward (white arrows in Fig. 2.2). Detailed FE-SEM observations revealed that the scale surfaces have unique periodic groove structures. Although the lengths of the groove structure periods are almost uniform within each scale, they seemed to vary between scales on the anterior regions of the body, particularly around the head (Fig. 2.2c). Generally, groove structure periods

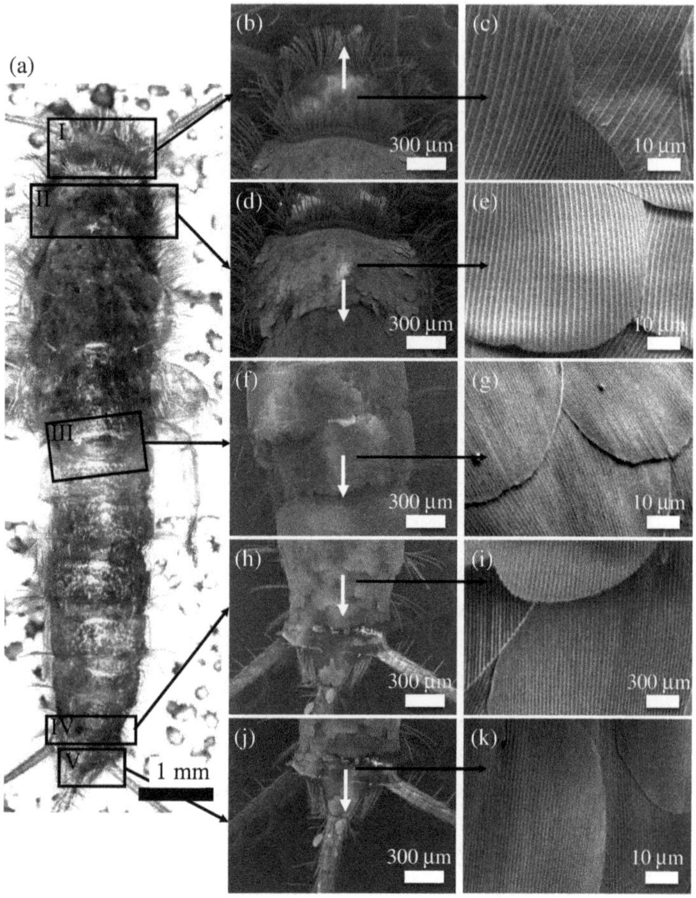

Figure 2.2 (a) Laser microscope image of a firebrat. (b-k) FE-SEM images of a firebrat surface.

of insect scales are uniform in each body spot, such as butterflies, whose scales have functions, including structural color [23]. Firebrats live in narrow spaces, so structural color as a scale function is unthinkable; their scales might be evolved for another function. To study the differences in groove structure periods more precisely, groove structure periods of each body position were analyzed. As a result, it was revealed that the groove structure periods and their standard deviations are clearly smaller as you move from the head to the tail. And also, lengths of groove structure periods become smaller toward the tail (Fig. 2.3). the back of the scale was also observed by

FE-SEM. Figure 2.4 shows FE-SEM images of a scale back and a cross section of a scale embedded in an epoxy resin. These images show that the back of the firebrat's scale is flat and groove structures are formed only the face side. This fact indicates that groove structures are not formed by buckling, as with wrinkles, and might be produced on purpose for some other function (one possibility of the necessity of a flat surface will be described later). Moreover, groove heights were analyzed to find common ground and strategies for evolution of scales. Figure 2.5 shows a graph of groove heights and distances from the scale root. In this case, six scales including three scales with a ~3.5 μm groove period (black) and three scales with a ~2.0 μm groove period (gray) were selected, respectively. According to the graph, groove heights are almost in proportion to distances from the scale root and regardless of groove periods. So groove heights are not constant; in other words, they vary with location.

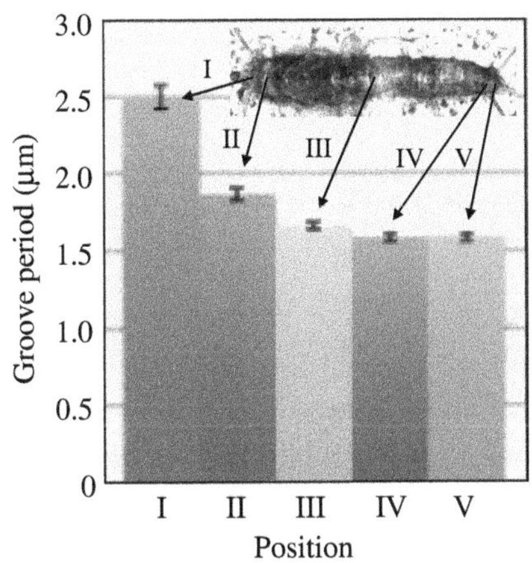

Figure 2.3 Graph of groove periods and standard deviations of each spots scales.

To summarize scale surface observations, the firebrat is covered by dense scales, and the scales have inhomogeneous groove periods on only the face side, especially around the head, which encounters the surrounded environment first. Groove heights become gradually higher from a scale root; groove heights are different from place to place by definition. For clearing these points, friction forces on the scale surfaces were measured using AFM in the next section, because the surface scales might have evolved for reducing friction to adapt their lifestyle.

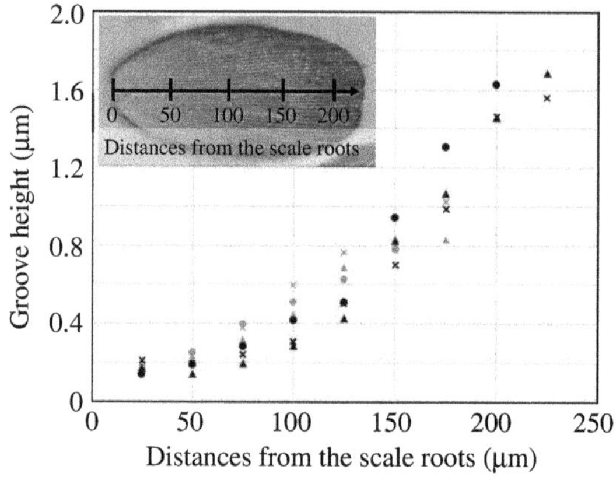

Figure 2.4 Graph of groove heights of scales.

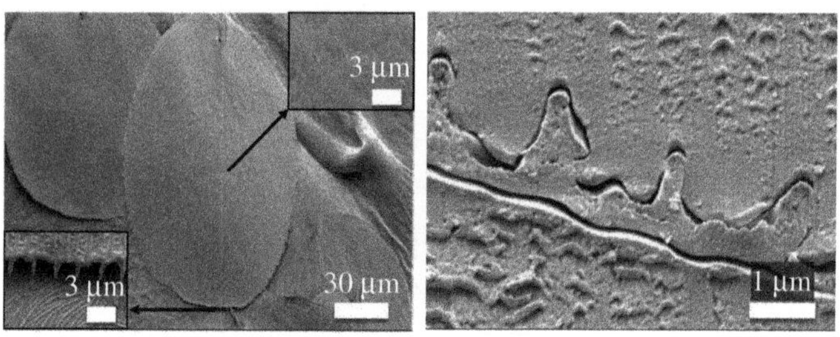

Figure 2.5 FE-SEM images of backsides and cross-section of scales.

2.3 Friction Property of Firebrat Scales

Surface friction forces were measured to study body surface friction properties, that is, scales, including scale boundaries. The firebrat was fixed on a silicon substrate by carbon adhesive tape for measurement of the horizontal area (Fig. 2.6). Then, friction forces of the firebrat surface were directly measured by AFM with a conventional needle-type probe (tip curvature radius = 8 nm) and colloidal probes with diameters of 5.0, 10, and 20 μm (material is silica). The scanning area was 50 μm square and the scan rate 0.3 Hz.

Scanning was performed in three directions: scale root to tip, scale tip to root, and lateral directions of scale groove structures.

Figure 2.6 Photograph of firebrat for measuring friction forces in case of including overlapped scales.

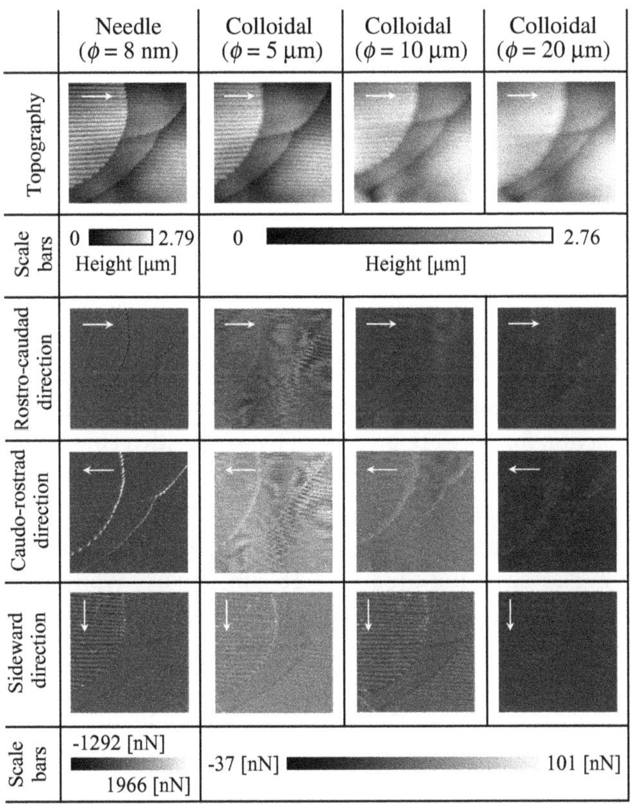

Figure 2.7 Topographies and friction force images of overlapped scales obtained by AFM measurements with a needle and colloidal probes.

Figure 2.7 shows friction force images on overlapped scales measured using AFM with three kinds of scanning directions. According to friction force images in Fig. 2.7, friction anisotropy at the boundaries between scales appeared in all probe cases. At the scale boundary, friction forces were markedly reduced when examined with scale-root-to-tip scanning. In contrast, friction forces were increased at the boundary with scale-tip-to-root scanning. This suggests that scale orientation on the trunk or thoracic and abdominal regions prevents sticking during forward movement into a narrow space. Meanwhile, firebrats might use high friction generated at the head scale tips (edges), whose growth direction is against the forward movement, for the mechanosensor to determine whether they can enter small spaces (the head is smaller than the trunk). Moreover, friction forces were increased at the top of the grooves in the case of lateral scanning. Groove structures appear to act as projections; however, this is not of concern for firebrats as they can not move sideways. Furthermore, when the probe diameter becomes larger, friction forces are decreased in all scanning directions. So contact areas between scale surfaces and probes might affect friction force; reduction of contact area reduces friction forces.

For investigation of the relationship between scale surface structures and dimensions of rubbed objects (indenters), a single scale with a ~2.0 μm groove period was taken from a firebrat and fixed on a silicon substrate using poly(vinyl alcohol) as an adhesive. Then, friction forces were measured by AFM with a common needle and colloidal probes with diameters of 2.0, 3.5, and 6.6 μm, which are almost the same or slightly larger than the groove periods (material is silica). Scanning was performed from scale root to tip only. The scanning area was 15 μm square, while the other measurement conditions were as described before. Figure 2.8 shows topographies and friction force images obtained by the friction force measurements. As a result of AFM measurements with three kinds of colloidal probes, the contact area decreased with increasing probe diameter in topographies. Because when indenters became larger than groove periods, the contact areas of the indenters formed a plane structure, negating the effects of the grooves. In the same manner, friction forces were reduced with increasing probe diameter. So friction force is proportional to the contact area; this result suggests that firebrats use scale groove structures for reduction of friction force. However, there are still questions as to why firebrat scale groove periods are uneven.

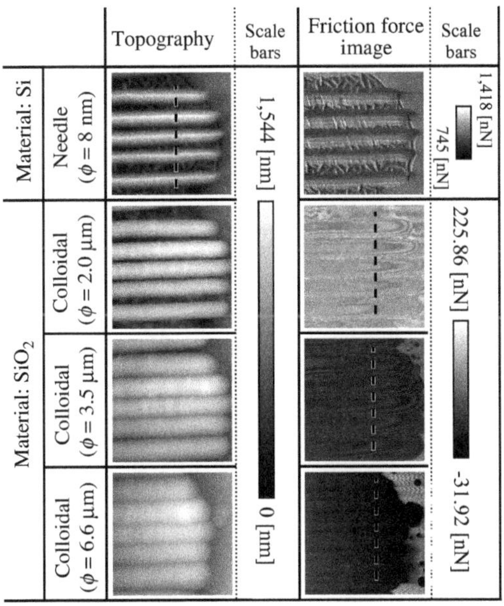

Figure 2.8 Topographies and friction force images of the scale obtained by AFM measurements with a needle and colloidal probes.

The relationship between groove structures and each colloidal probe are compared in Fig. 2.9, which shows the height profile of groove structures (gray line) obtained from the topography and friction forces (black lines, including dotted and broken lines) obtained from the friction force images in Fig. 2.8 (selected positions are displayed by black broken lines). According to the graph, friction forces are remarkably large at the bottom of the grooves in the case of the 2.0-μm-diameter probe. In this experiment, the scale groove period was ~2.0 μm, so it is almost the same size as the probe diameter. To stick the probe in grooves, friction forces become large. On the other hand, probes were in contact just the top of the grooves, and friction forces were lower in the case of 3.5- and 6.6-μm-diameter probes. These facts clearly suggest that an increment in contact area causes larger friction forces. In particular, if the probe diameter and groove periods are of the same size, the probe sticks in the grooves, leading to a larger friction force. According to these results, groove structures are definitely thought to influence friction, and inhomogeneous groove periods may prevent high levels of friction with their surrounded rough surfaces of specific size by preventing fixation of all the scales. This may be one reason why firebrats intentionally vary groove periods, especially around the head, which frequently comes into contact with their surroundings, whereas standard deviations of the scales on parts that have little contact, such as the posterior abdomen or tail, are small.

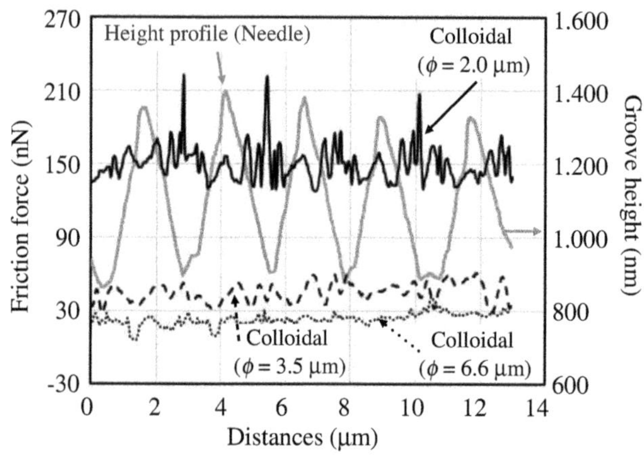

Figure 2.9 Graphs of a groove height profile and friction forces of the scale obtained by AFM measurements with a needle and colloidal probes. Selected positions of that were displayed in Fig. 2.7 as black broken lines.

2.4 Summary

In this chapter, the body surface of firebrats was observed by FE-SEM and friction force measured by colloidal probe AFM as part of investigations of the properties of firebrat scales. FE-SEM observations revealed that firebrats are covered in dense scales, equipped with microgrooves. Detailed observations and analyses indicate that the groove periods vary by body part, although the groove periods are almost uniform within each scale. Friction measurements suggest that firebrat scales might have five properties; one is the scale itself. Though we have not mentioned this before, when a scale is injured or trapped by something, including a spider's net, the firebrat removes the scale to escape (scales are regenerated by molting). The second is the direction of scale growth on the trunk or thoracic and abdominal regions. This prevents the firebrat from becoming lodged during movement in narrow spaces. The third is the growth direction of the head scales. The head scales act as a mechanosensor to identify whether the insect can enter a narrow space. The fourth is groove structures, which reduce the contact area and lead to reduction of friction forces. It is also suggested that scale backs are flat. To form flat backs of scales, the contact area and friction force between scales are also reduced, and this prevents sticking of scales to each other.

The final property is the heterogeneous groove periods. Due to changes in the groove period for each scale, it is difficult to fix the whole-body scales, particularly on rough surfaces. If some scales are stuck to something, the firebrat can easily escape by removing only the trapped scales. This is also suggested by the standard deviations of groove periods, which are larger toward the anterior portion of the body, which is frequently in contact with the environment, and smaller on the body parts, such as the tail, that have little external contact. This research shows that firebrats adapt their scales to their habitat. On the basis of this knowledge of firebrat scales, we can develop novel, less frictional surfaces applicable to a wide variety of industrial fields.

Acknowledgments

This work was supported by the Japanese Ministry of Education, Culture, Sports, Science and Technology (MEXT) via Grants-in-Aid for Scientific Research on Innovative Areas (grant no. 24120003) and a Grant-in-Aid for Young Scientists A (grant no. 26712028). Firebrats and their breeding methods were explained and lectured by Prof. R. Machida, Sugadaira Montane Research Center, University of Tsukuba.

References

1. Nakada, M. (1994). Trends in engine technology and tribology, *Tribol. Int.*, **27**(1), pp. 3–8.

2. Gong, J., Iwasaki, Y., Osada, Y., Kurihara, K. and Hamai, Y. (1999). Friction of gels. 3. Friction on solid surfaces, *J. Phys. Chem. B*, **103**(29), pp. 6001–6006.
3. Hörner, D. (2002). Recent trends in environmentally friendly lubricants, *J. Synthetic Lubric.*, **18**(4), pp. 327–347.
4. Miyake, K., Nakano, M., Korenaga, A., Mano, H. and Ando, Y. (2010). Tribological properties of nanostripe surface structures-a design concept for improving tribological properties, *J. Phys. D Appl. Phys.*, **43**(46).
5. Sasaki, S., Ogawa, H., Korenaga, A., Miyake, K., Nakano, M. and Murakami, T. (2010). Effects of surface texture size on the tribological properties of slideways, *P. I. Mech. Eng. J.-J. Eng.*, **224**(J9), pp. 885–890.
6. Wang, X., Kato, K., Adachi, K. and Aizawa, K. (2003). Loads carrying capacity map for the surface texture design of SiC thrust bearing sliding in water, *Tribol. Int.*, **36**(3), pp. 189–197.
7. Kobayashi, M., Terayama, Y., Hosaka, N., Kaido, M., Suzuki, A., Yamada, N., Torikai, N., Ishihara, K. and Takahara, A. (2007). Friction behavior of high-density poly(2-methacryloyloxyethyl phosphorylcholine) brush in aqueous media, *Soft Matter*, **3**(6), pp. 740–746.
8. Overney, R. M., Takano, H., Fujihira, M., Meyer, E. and Güntherodt, H. J. (1994). Wear, friction and sliding speed correlations on Langmuir-Blodgett films observed by atomic force microscopy, *Thin Solid Films*, **240**(1,Äì2), pp. 105–109.
9. Baio, J. E., Spinner, M., Jaye, C., Fischer, D. A., Gorb, S. N. and Weidner, T. (2015). Evidence of a molecular boundary lubricant at snakeskin surfaces, *J. R. Soc. Interface*, **12**(113).
10. Baum, M. J., Heepe, L. and Gorb, S. N. (2014). Friction behavior of a microstructured polymer surface inspired by snake skin, *Beilstein J. Nanotechnol.*, **5**, pp. 83–97.
11. Berthe, R. A., Westhoff, G., Bleckmann, H. and Gorb, S. N. (2009). Surface structure and frictional properties of the skin of the Amazon tree boa Corallus hortulanus (Squamata, Boidae), *J. Compar. Physiol. a-Neuroethol. Sensory Neural Behav. Physiol.*, **195**(3), pp. 311–318.
12. Bhushan, B. (2009). Biomimetics: lessons from nature; an overview, *Philosoph. Trans. R. Soc. a-Math. Phys. Eng. Sci.*, **367**(1893), pp. 1445–1486.
13. Autumn, K., Liang, Y. A., Hsieh, S. T., Zesch, W., Chan, W. P., Kenny, T. W., Fearing, R. and Full, R. J. (2000). Adhesive force of a single gecko foot-hair, *Nature*, **405**(6787), pp. 681–685.

14. Voigt, D., Schweikart, A., Fery, A. and Gorb, S. (2012). Leaf beetle attachment on wrinkles: isotropic friction on anisotropic surfaces, *J. Exp. Biol.*, **215**(11), pp. 1975.

15. Arzt, E., Gorb, S. and Spolenak, R. (2003). From micro to nano contacts in biological attachment devices, *Proc. Natl. Acad. Sci.*, **100**(19), pp. 10603–10606.

16. Varenberg, M. and Gorb, S. N. (2009). Hexagonal surface micropattern for dry and wet friction, *Adv. Mater.*, **21**(4), pp. 483–486.

17. Ohde, T., Masumoto, M., Yaginuma, T. and Niimi, T. (2009). Embryonic RNAi analysis in the firebrat, *Thermobia domestica*: Distal-less is required to form caudal filament, *J. Insect Biotechnol. Sericol.*, **78**(2), pp. 2_99–92_105.

18. H. L. Sweetman (1938) Physical ecology of the firebrat, *Thermobia domestica* (Packard), *Ecol. Monogr.*, **8**(2), pp. 285–311.

19. Watson, J. A. L. (1964). Moulting and reproduction in the adult firebrat, *Thermobia domestica* (Packard) (Thysanura, Lepismatidae)—II. The reproductive cycles, *J. Insect Physiol.*, **10**(2), pp. 399–408.

20. Ducker, W. A., Senden, T. J. and Pashley, R. M. (1991). Direct measurement of colloidal forces using an atomic force microscope, *Nature*, **353**(6341), pp. 239–241.

21. Francius, G., Hemmerle, J., Ohayon, J., Schaaf, P., Voegel, J. C., Picart, C. and Senger, B. (2006). Effect of crosslinking on the elasticity of polyelectrolyte multilayer films measured by colloidal probe AFM, *Microsc. Res. Tech.*, **69**(2), pp. 84–92.

22. Zauscher, S. and Klingenberg, D. J. (2001). Friction between cellulose surfaces measured with colloidal probe microscopy, *Colloids Surf. A: Physicochem. Eng. Aspects*, **178**(1,Äì3), pp. 213–229.

23. Kinoshita, S., Yoshioka, S. and Kawagoe, K. (2002). Mechanisms of structural colour in the Morpho butterfly: cooperation of regularity and irregularity in an iridescent scale, *Proc. R. Soc. Lond. B: Biol. Sci.*, **269**(1499), pp. 1417–1421.

Chapter 3

Biomechanics and Biomimetics in Flying and Swimming

Hao Liu,[a,b] **Toshiyuki Nakata,**[a] **Gen Li,**[c] **and Dmitry Kolomenskiy**[c]

[a]*Graduate School of Engineering, Chiba University, 1-33 Yayoi-cho, Inage-ku, Chiba, China*
[b]*Shanghai-Jiao Tong University and Chiba University International Cooperative Research Centre (SJTU-CU ICRC), 800 Dongchuan, Minhang, Shanghai, China*
[c]*Japan Agency for Marine-Earth Science and Technology, 3173-25, Showa-machi, Kanazawa-ku, Yokohama-city, Kanagawa, Japan*
hliu@faculty.chiba-u.jp

Flying and swimming in nature present sophisticated and exciting ventures in biomimetics, which seeks sustainable solutions and solves practical problems by emulating nature's time-tested patterns, functions, and strategies. Biomechanics in biological flying and swimming is an interdisciplinary field in exploring essential novel mechanisms that have been validated through a long period of natural selection, which is expected to offer an alternative paradigm for industrial biomimetic designs. Biofluid dynamics, a major research area in biomechanics associated with insect and bird flights and fish swimming, is of high dynamics and complexity

Industrial Biomimetics
Edited by Akihiro Miyauchi and Masatsugu Shimomura

ISBN 978-981-4800-07-5 (Hardcover), 978-0-429-05883-7 (eBook)
www.jennystanford.com

and normally brings a low-Reynolds-number (Re) regime and high unsteadiness challenges in modeling and analysis of the physics of flows. In this chapter, we intend to overview recent advances in unsteady biofluid dynamics in terms of leading-edge vortices, passive mechanisms in flexible wings and hinges, and flapping flight in unsteady environments, as well as undulatory swimming, flapping-fin hydrodynamics, body–fin interactions, C-start and maneuvering, swimming in turbulence, and collective swimming. We further highlight the state of the art in biomimetics of flapping-wing micro–air vehicles (MAVs) and swimming robotics MAVs. We conclude with a perspective outlook on future challenges and tasks of several key issues of the field.

3.1 Introduction

Motivated by the emerging field of biomimetics, in particular biorobotics of insect and bird size drones—flapping micro–air vehicles (MAVs) [123] that can perform autonomous flight in a natural environment—researchers from various fields of mathematics, biology, mechanical and aeronautical engineering, computer engineering, etc., have made tremendous efforts in unveiling the unsteady biomechanical mechanisms in flapping-wing flight. Biomimetics that seeks sustainable solutions and solves practical problems by emulating nature's time-tested patterns, functions, and strategies has shown remarkable development in recent years [107]. Now it has been widely recognized that flying and swimming in nature present sophisticated and exciting ventures in biomimetics and are expected to inspire and innovate new technologies in biorobotics [123].

A common feature associated with animal flying and swimming is that biological flyers and swimmers propel and control themselves in air or water through oscillating their wings, fins, and bodies. Therefore, to unveil the novel mechanisms of biological flying and swimming, we need to answer a central question on the physics of fluid structure (wings, fins, and bodies) interaction (FSI) under various unsteady environments. Given the complex geometries and highly unsteady motions of flapping wings and fins as well as undulating bodies, mathematical modeling of the FSI problems in concern with flying aerodynamics and swimming hydrodynamics brings great challenges in terms of a low-Reynolds-number (Re) regime and high unsteadiness. As a simplified physical model, biofluid dynamics in flying and swimming have been until now mostly studied with a specific focus on the flow phenomena associated with a body or wings moving in a steady flow [200].

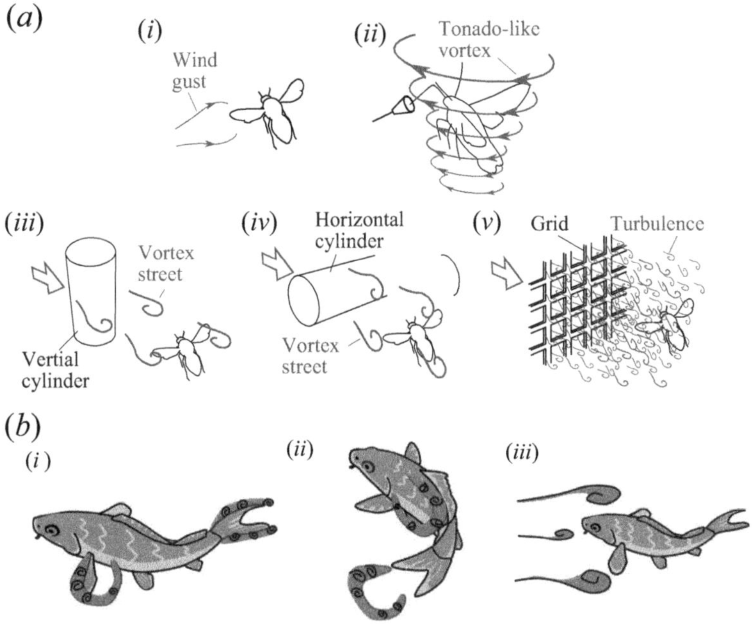

Figure 3.1 Fishes, insects, and birds perform complex maneuvers while coping with unsteady environments. (a) Unsteady ambient air flow conditions realized in laboratory experiments with flying animals: (i) wind gust, (ii) vertical tornado-like vortex, (iii) vortex street past a vertical cylinder, (iv) vortex street past a horizontal cylinder, and (v) grid turbulence. (b) Unsteady hydrodynamics in swimming is distinguished into three categories: (i) unsteady hydrodynamics in steady swimming, (ii) hydrodynamics in C-start and maneuvering, and (iii) swimming hydrodynamics in unsteady environments.

In reality, characterized by unsteady wing flapping and body undulation, FSIs, flexible wings and bodies, and turbulent environments, as well as complex maneuvers (Fig. 3.1), biofluid dynamics in insect and bird flights as well as in fish swimming are highly dynamic and unsteady, complex, and even *chaotic* in turbulent environments [80]. Therefore, with the unsteadiness factor, the FSI problem in flying and swimming encounters big challenges in physical and mathematical modeling and computational approach of the biofluid dynamics. In the past decades, there has been a significant increase of interests and works on exploring the unsteady mechanisms of biofluid dynamics, for instance, as illustrated in Fig. 3.1, in insect and bird flights (Fig. 3.1a), and in fish swimming (Fig. 3.1b). Here, we give a topical overview on recent advances in unsteady biofluid dynamics and then conclude with a perspective view on several key topics of the field.

3.2 Unsteady Biomechanical Mechanisms in Flapping Flight

Flying insects and birds that power and control flight by flapping their wings are capable of performing excellent flight stability and complex maneuvers under various unsteady environments, while steering and maneuvering by rapidly and continuously varying their wing kinematics [123]. Flapping-wing aerodynamics in biological flights is normally characterized by large-scale vortex structures, complicated flapping-wing kinematics, and flexible wing structures; the flow past their flapping wings is intrinsically unsteady and highly 3D and even turbulent. Therefore a challenging problem in uncovering aerodynamic mechanisms in flapping flight is to answer a central question on the generation of the large-scale vortex dynamics and, hence, as well as their correlation with force production.

3.2.1 Leading-Edge Vortices of Flapping and Revolving Wings

Earlier studies have identified multiple aerodynamic mechanisms that are responsible for enhancing the aerodynamic force production in flapping wings.

The leading-edge vortex (LEV) is perhaps the most robust one among them, which has been confirmed in experiments with animals and plants, dynamically scaled mechanical models, and numerical simulations. Earlier literature survey can be found, for example, in Ref. [166], Section 4.4.1.

Small biological fliers have thin wings. Insects, for example, typically have membrane wings with stiff veins at the sharp leading edge. Hence, the flow separates from the leading edge and forms a strong vortex that persists above the suction side of the wing during most parts of the flapping stroke; see a sample flow visualization in Fig. 3.2a. A pressure deficit in the vortex core extends to the wing surface, leaving a triangular imprint of low pressure in the region adjacent to the leading edge. The aerodynamic force is enhanced as a consequence, typically showing the sectional lift coefficient values of 2 or greater [100]. This effect is referred to as *delayed stall*.

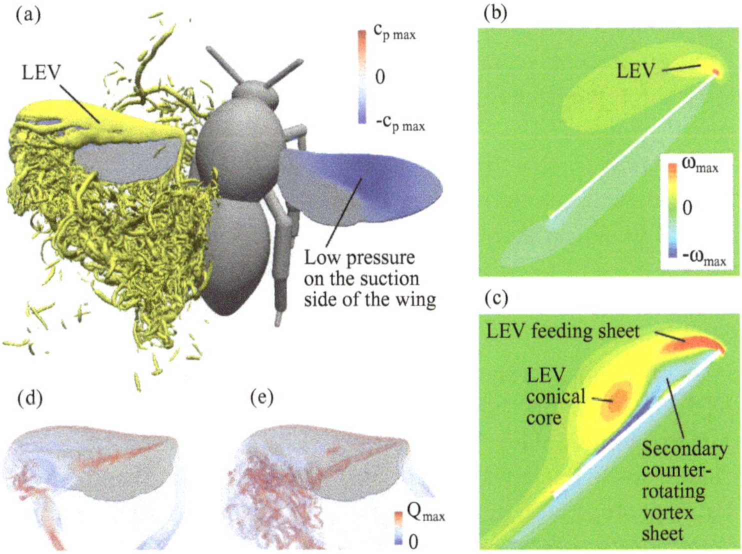

Figure 3.2 Sample visualization of the LEVs over flapping and revolving wings, obtained using numerical simulation. (a) Q-criterion iso-surface over the left wing of a hovering bumblebee (yellow color) and pressure distribution over the wing and body surface (blue-red color scale). (b) Sectional plot of the spanwise vorticity component of a revolving fruit fly wing at Re = 100. (c) Sectional plot of the spanwise vorticity component of a revolving bumblebee forewing at Re = 2100. (d) Volume visualization of the Q-criterion in the flow over a revolving bumblebee wing at the onset of the LEV breakdown (rotation angle $\varphi = 81°$) and (e) long after the LEV breakdown ($\varphi = 138°$) at Re = 3000.

Although the LEV has been recognized as a convergent feature common to many flying animals, the physical effects and parameters that control its attachment have been debated over the past decades. Recent particle image velocimetry (PIV) experiments, time-resolved force measurements, and high-fidelity numerical simulations have helped to provide definite answers to many of the open questions. Studies using flapping (e.g., [121, 106]) or simple revolving (e.g., [27, 145]) kinematic protocols and realistic wing planforms with flat profile (e.g., [121, 106]), cambered profile [72], or rectangular wing shapes (e.g., [145, 86]) have allowed us to identify the general aspects of the LEV dynamics.

One of the important tools in this field of research is the experiment with a dynamically scaled-up robotic wing. With advanced PIV systems and mechanical devices, more accurate and detailed information of near- and far-field flow structures has been acquired. Recent experiments include stereoscopic PIV (e.g., [86, 147]) or tomographic PIV [34, 145], in combination with force measurements [145, 27], which are capable of correlating between the LEV flow features and force production.

Numerical simulation has also proved to be an effective tool [122, 120]. A persistent increase in computer performance has now made it possible to examine the LEV in a wide range of flow regimes, from low-Reynolds-number laminar flow to turbulent flow at an Re of the order 10,000. The Reynolds number of a flapping wing is normally defined as the ratio of the mean wing tip speed multiplied by the mean chord length, divided by the kinematic viscosity of the surrounding fluid. Liu and Aono [121] studied the size effects from thrips to the hawkmoth in a biologically accurate setup, that is, flapping wings with a morphology and kinematic protocol derived from measurements made on the respective insects. Harbig [73] examined an insect wing model with a biologically realistic contour following a simplified revolving kinematic protocol.

The aspect ratio (AR) of the wing varied with rescaling of its contour in the spanwise direction. Garmann and Visbal [64] conducted implicit large eddy simulations (ILES) of revolving rectangular wings. Kolomenskiy et al. [98] considered revolving triangular wings with a constant relative local chord length c normalized by the distance from the rotation axis r.

From a theoretical perspective, Maxworthy [132] emphasized the importance of spanwise flow above the suction surface of the wing and proposed a model to associate it with the pressure gradient due to centrifugal force as well as with the spanwise variations in vortex size and circulation. He further pointed out to the similarity with the LEVs of swept delta wings. Lentink and Dickinson [105] noticed that a combined contribution of the centrifugal and Coriolis terms in the momentum equation relative to the advection term scales in an inverse proportion with the wing tip distance from the rotation axis, and in a direct proportion with the wing chord length. Limacher et al. [117] derived a reduced-order model for the axial streamline of a stable LEV and showed that the Coriolis acceleration can only have a destabilizing effect by tilting the LEV core toward the trailing edge. Chen et al. [30] recently derived an analytical solution for the edge vortex of a revolving rectangular plate, which explained the growth of the vortex core in the spanwise direction by a balance between vorticity production at the leading edge and its transport toward the wing tip.

In view of the remarkable variability in the size of animals that use flapping wings for flight, a practically important question is the LEV sensitivity to variation of the Reynolds number or sizing. The majority of studies focus on the range of the Re from one thousand to several thousand (e.g., [27, 147]). The results presented in Refs. [106, 143, 100] extend it up to Re = 25,000, 14,000, and 21,600, respectively.

Analysis of the flow at Re = ~100–200 can be found in, for example, Refs. [98, 106]. Finally, a broad range of Re from 10 to 6300 is covered in Ref. [121] and from 200 to 60,000 in Ref. [65]. These and earlier studies provide solid evidence that the LEV structure varies remarkably from a diffuse vortex at Re of the order 100 to a compact core when Re is above 1000.

Typical sectional vorticity plots are shown in Fig. 3.2b,c. The spanwise component of the vorticity at Re = 100 (Fig. 3.2b) has its maximum at the leading edge. A shear layer develops between the free stream above the wing and the recirculation bubble immediately behind the wing. The vorticity sheet produced at the leading edge diffuses rapidly.

At larger Reynolds numbers, above 1000, the vorticity sheet gets thinner, and it feeds a conical vortex core (see Fig. 3.2c). Thus, there appear two local maxima of the vorticity: at the leading edge and in the core. This structure is referred to as an arch-like LEV [64, 27]. As it remains close to the suction surface of the wing, it induces an opposite-sign vorticity layer that can also separate, if the Reynolds number is sufficiently large, and form a secondary counterrotating vortex. Wojcik and Buchholtz [196] pointed out a large vorticity cancellation between the negative secondary vortex and the positive primary LEV core. Note, however, that both cores contribute to the regulation of vorticity production at the leading edge. Therefore, the vorticity cancellation should be accompanied by enhanced vorticity production at the edge.

The role of the Coriolis force has been addressed in several recent studies. In the reference frame rotating with the wing, its azimuthal and radial components are nonzero and its prevailing direction is away from the wing surface toward the wake and inward toward the rotation axis [65, 64]. Limacher et al. [117] showed that the Coriolis force has a destabilizing effect by curving the neutral streamline toward the trailing edge and hence approaching the shedding. However, the LEV core also expands as a result of vorticity production at the leading edge [30], and this expansion may be more significant.

In flapping flight, since the amplitude of one half-stroke is less than 180 degrees, the LEV can hardly reach its steady state. Experiments by Philips et al. [146, 147] using a dynamically scaled flapping mechanical apparatus show qualitatively the same phenomena as found for revolving wings, but the LEV does not reach its fully developed state before the reversal. In hovering flapping flight near the ground, the LEV remains attached to the wing even when the wing is closest to the ground, but its circulation varies with the ground clearance [127]. Additional complication arises in forward flight, when the uniform inflow velocity is superposed with the flapping velocity. The LEV stability in that case depends on the advance ratio, defined as the ratio of the flight speed to the average velocity at the wing tip due to its rotation [74].

The fundamental progress that has been made during the past decade opens new perspective topics for future research such as theoretical estimation of the LEV-induced aerodynamic force, reduced-order modeling of the LEV dynamics, and optimal design to maximize the aerodynamic benefits of the LEVs.

3.2.2 Passive Mechanisms in Flexible Wings and Hinges

Insects' flapping wings change their 3D shapes passively and dynamically in response to aerodynamic and inertial forces [198], generating unsteady aerodynamic forces to stay airborne. The flapping wings must be light to reduce the mechanical power to overcome its inertia; therefore, insect wings are inevitably thin and flexible. When modeling, the instantaneous wing shape must be identified iteratively since the wing kinematics and the flow structure determine the change of wing shapes, while the wing deformation affects its aerodynamics (Fig. 3.3a). Such FSI between flexible wings and surrounding fluid is of great importance and may significantly affect the aerodynamics, flight control, and maneuverability of a flyer, since it forms a mechanical feedback loop with which to modulate the wing shape and kinematics passively, in addition to an active sensory feedback loop [48, 123].

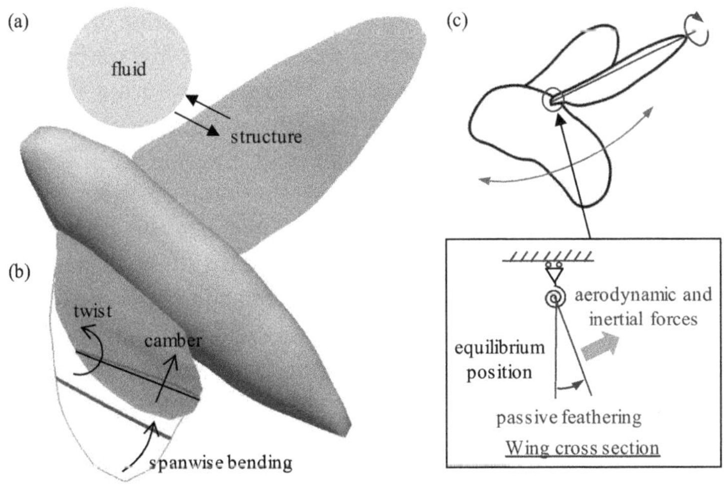

Figure 3.3 (a) Interaction between the surrounding fluid and flexible wings results in wing deformation, such as (b) twist, camber, and spanwise bending. (c) The wing hinge is also flexible and can modulate the angle of attack (AoA) passively.

Insect wings consist of multiscale complex structures [199]. The wing veins mainly determine the wing planform and inherent out-of-plane shape, such as corrugation [155], together with thin membranes. The structure of the insect wing, such as the pattern of venation and microstructure, passively determines the aerodynamics of flexible flapping wings because, unlike other flying animals such as birds or bats, insect wings do not have musculature inside. The veins run from the wing base to the distal part or trailing edge of the wing, tapering toward the tip. Such typical venation pattern causes the anisotropy in spanwise and chordwise bending stiffness ranging from 1 to 2 orders of magnitude [37] and the distribution of the stiffness from a relatively stiff base to a soft tip [36, 104, 171]. Some veins are jointed by soft rubberlike protein, called resilin [7, 51, 193]. The resilin in the hind wings of beetles distributes to add extra flexibility on the wings for folding [71]. In bumblebee wings, stiffening one single joint can increase the flexural stiffness by about 40% [134].

The interaction between inertial and aerodynamic forces together with the wing structure determines the wing deformation and hence the aerodynamic performance of flexible flapping wings. Combes and Daniel [37] measured the wing deformation of a rotated hawkmoth wing in air and helium and suggested the low contribution of aerodynamic forces to the wing deformation during flapping. Jongerius and Lentink [87] found, through computational structural dynamics and a quasi-steady model, that the inertial forces on dragonfly wings could be three times higher than the aerodynamic forces. However, Zheng et al. [207] pointed out the importance of aerodynamic forces in a butterfly's forward flight through the measurement of wing deformation. Sims et al. [167] measured the frequency response of hawkmoth wings in vacuum and air and found a frequency shift of 26.5% from vacuum. These results indicate the inertial force on flapping wings as the major determinant of flapping wing deformation, but the aeroelastic contribution is not negligible when it comes to assessing the dynamic response of flexible flapping wings accurately.

One critical value concerning the flexible wing hinge is the resonance, since flapping wings may be able to store and release the energy for flapping motion, as suggested in insect muscles [49]. The resonance of insect flapping wings and its ratio to the wingbeat frequency are measured in various insects. Most of the recent studies suggested a low wingbeat frequency compared to the natural frequency [70, 31, 172]. The wingbeat frequency of dragonflies are merely 16% of the natural frequency [31]; Ha et al. [70] measured the natural frequency of the flapping wings of eight insect species and confirmed the lower wingbeat frequency than the natural frequency except for Hymenoptera. Lehmann et al. [104] investigated the wing kinematics and wing structure and concluded that the kinetic energy that is stored in the form of elastic energy on the wing is small and contributes slightly to the recycling of energy at the next half stroke. The amplitude of the oscillation can be increased without increasing the input force at resonance. However, the aerodynamic performance of chordwise flexible wings is suggested to be maximized at a lower frequency than the resonance because of the importance of the phase lag of wing rotation with respect to the wing translation (e.g., [149, 169]). The observed natural frequency shown above suggests that insects may utilize the enhancement of aerodynamic performance by the phase lag between wing rotation and translation, rather than matching the natural frequency and wingbeat frequency. Scaling laws are extremely useful to describe the complex phenomena like FSI of flexible flapping wings. The Reynolds number and Strouhal number, or reduced frequency, are well-known examples to describe aerodynamics and unsteadiness [164]. Some scaling laws for force generation and efficiency have been suggested using a simplified model of a flapping wing in air or water (e.g., [88, 165]).

The passive, dynamic, and 3D shape change in insect flapping wings has a specific impact on aerodynamic performance. In the bumblebee, stiffening the resilin joint on the wing can reduce the maximum force production [134] and flight stability [133]. Several experimental studies confirmed that passive wing deformation could be at the same level or reduce the performance compared to rigid wings [128, 206]. Inertial force on a flapping wing is highest at the stroke reversal when the wing acceleration is maximized, while the aerodynamic force is highest at the midstroke when the wing velocity is maximized. The phase difference due to the balance of such inertial and aerodynamic forces may affect the aerodynamic performance. Yin and Luo [202] performed computational studies with artificial structural models and found both inertia- and fluid-induced wing deformations can enhance the lift force but fluid-induced wing deformation is more efficient. Eldridge et al. [54] found the reduction of the sensitivity of lift generation to the phase of input rotation. The wingbeat frequency that is lower than the natural frequency of the wing implies the quick aeroelastic response of the flexible wing against the unsteady aerodynamics on the wing, such as the development and breakdown on the wing. Computational studies based on a realistic structural model of the hawkmoth's flexible wings found that the lift force and efficiency can be increased by delayed stall of the LEV on the flexible flapping or revolving wings and the passive adjustment of the phase and speed of rotation at around the distal part of the wing [136, 138]. The passive deformation may also work for flow control under the unpredictable gust or turbulence that is suggested for the wings of a steppe eagle [28], but more studies with experiments and computation are necessary to conclude the effectiveness of wing flexibility for gust mitigation. Photogrammetric reconstruction helps to identify dynamic shape changes, such as twists, cambers, and spanwise bending (Fig. 3.3b), of flapping wings in various insects [187, 207] and the aerodynamic consequences. A comparison of aerodynamic performance between measured and artificial wing kinematics confirmed that the twist could enhance the efficiency of locusts, butterflies, and beetles in forward flight [203, 207, 103]; the camber might not do at high angles of attack corresponding to realistic flapping wings [181, 53]. Harbig et al. [72] confirmed, however, the increase in the lift-to-drag ratio with positive camber by tilting the force normal to the wing surface around the leading edge at higher angles of attack.

While the wing kinematics in flying insects is controlled at the wing hinge after the transmission of forces from power and steering muscles, the wing hinge is also known to possess flexibility [59] (Fig. 3.3c). It is shown through analyses of the aerodynamic and inertial power that the wing rotation measured in insects can be achieved passively at the wing hinge [17]. The wing rotation around the spanwise axis has a drastic impact on the aerodynamic performance of flapping wings since the rotation directly affects the angle of attack (AoA) throughout the wingspan. The interplay among the flexibility at the wing hinge, wing inertia, and Reynolds number effect determines the dynamics of the rotation and lift generation capability [82, 83, 81]. The active modulation of the flexibility at the wing hinge is suggested to enable the changes in the AoA during fruit flies' maneuvers even with such flexible wing hinges [16, 14]. The idea of passive modulation of the AoA can greatly simplify the flapping-wing mechanism and has been employed for bioinspired MAVs [195, 32].

The recent progress summarized here has been made by the advancement of technologies for photogrammetry and computational modeling. Computational FSI analyses are, however, generally computationally expensive, and it is still not practical to use the simulators for designing and optimizing the flexible flapping-wing structure. Therefore, it is necessary to develop a more computationally cheap theoretical model for a 3D flapping wing. An experimental model would also help to advance our understanding of the flexible wing aerodynamics. It is challenging to emulate and fabricate biological flexible structure, but new technologies such as microelectromechanical systems (MEMS) and 3D printing can be utilized for the development of a biomimetic flexible structure.

3.2.3 Dynamics of Animal Flight in Turbulence

A new topic of research that emerged recently is the flight dynamics of animals in an unsteady air environment. Atmospheric flows are usually turbulent; the turbulence intensity and scale may be relatively large compared to the flying animals' speed and size. Hence, understanding of the mechanisms that the flyers employ to deal with atmospheric turbulence is important for placing their flight dynamics in the ecological context, as well as for exploring robust turbulence rejection mechanisms in bioinspired flight systems for MAV applications [123].

The aerial environment of flying animals is complex and variable, as it incorporates unsteady flows of diverse origins, such as large-scale wind gusts, instability of the inflected mean flow velocity profiles in vegetation canopy layers, small-scale turbulence produced in the wakes of the foliage, etc. [61]. The impact of wind conditions on animal flight has been studied in relation to migration [29] and foraging [40] using field observations. Furthermore, laboratory experiments have been set up to study the flight dynamics at the level of an individual animal under controlled unsteady flow conditions. Combes and Dudley [38] observed for the first time the free flight of orchid bees in a turbulent air jet. They found that the maximum speed that the bees could maintain in the turbulent jet was less than normally achieved in quiescent air, and it was limited by body roll oscillations that increased with the jet speed. The bees attempted to stabilize their orientation by extending their hind legs at higher flight speeds.

Vance et al. [183] carried out a comparative study on the sensitivity of honeybees and stalk-eye flies to localized wind gusts (see a schematic drawing in Fig. 3.1a-i). The typical flight of both species included a passive perturbation period, when the insect body orientation was destabilized by the gust, followed by an active perturbation period characterized by compensatory modulation of wing kinematics, and then active and passive recovery, when the insect restored its attitude after the gust stopped. However, it was found that bees and stalk-eye flies respond differently to aerial perturbations. Bees show significant roll rotations to which they respond by bilateral asymmetry in stroke amplitude. Stalk-eye flies show greater yaw instability during the perturbation period and use a combination of asymmetric stroke amplitude and wing-feathering rotation angle as a compensatory mechanism. The asymmetry in wing-feathering rotations, which can directly control the kinematic AoA, was also suggested by Ristroph et al. [156] as the aerodynamic mechanism used by fruit flies to recover form external perturbation in free flight. In the latter study, free flight was impulsively perturbed by applying an external magnetic field on ferromagnetic pins glued to the flies.

The effect of sustained unilateral aerodynamic perturbation has been explored using experiments with hawkmoths in feeding flight. Ortega-Jimenez et al. [141] studied hawkmoth flight in a vertical tornado-like vortex (see Fig. 3.1a-ii). It was found that the hawkmoths change yaw orientation, compared to hovering in quiescent air, which is explained by the drag difference between the left and right wings. The study also reported significant changes in the wing stroke plane angle, wingbeat amplitude, and AoA, which generated an aerodynamic torque opposite to that induced by the tornado vortex.

Flight in a vortex street behind a cylinder has received great attention as a biologically relevant case realizable in the laboratory (see Fig. 3.1a-iii,iv). Ravi et al. [151] reported that bees are most sensitive to the lateral perturbations induced by a cylinder positioned vertically, displaying large rolling motions. Ravi et al. [153] showed that bees do not attempt to mitigate the high-frequency rolling motion but control their body attitude at a larger timescale commensurate with the frequency of their usual casting motion. For a hawkmoth in feeding flights in vortex streets past vertical cylinders of different size, Ortega-Jimenez et al. [140] reported destabilizing effects on both yaw and roll, as well as an increase in wingbeat frequency in some cases. Similar experiments with hummingbirds showed increases in variances of wingbeat frequency, body roll, pitch, and yaw amplitudes when the cylinder was large enough. The metabolic rate under increased by up to 25% compared to flights in a laminar inflow at the same airspeed [142, 139].

Experiments with bumblebees flying in a grid turbulence have been conducted by Crall et al. [40] (see Fig. 3.1a-v). When the airspeed was small (1.5 m/s), no significant changes in the wing kinematics were detected. At a larger airspeed (3 m/s), the wingbeat frequency and the stroke amplitude in the turbulent flow trial became larger than in the laminar case. In addition, the bees employed stroke amplitude asymmetry, presumably attempting to generate aerodynamic torque reducing the roll, as the stroke amplitude asymmetry was correlated with the roll amplitude. However, the compensation was not perfect, as the residual roll was significantly larger in turbulence than in the laminar flow.

Another study on hummingbirds flying in a grid turbulence by Ravi et al. [152] showed both steady and unsteady changes in several kinematic variables in response to perturbations imposed by the turbulent flow. The flapping frequency, mean stroke amplitude, anatomical stroke plane angle, and tail-fanning angle increased in order to produce larger aerodynamic forces and/or improve passive stability. The hummingbirds also displayed increasing stroke-to-stroke variability in these kinematic parameters, as well as bilateral asymmetry in the stroke amplitude and the stroke plane angle, when flying under highly turbulent conditions. Despite the significant variability of wing kinematics and unsteadiness of the body orientation, the hummingbirds were capable of maintaining remarkably constant head position relative to the feeder.

There are potential energetic and control limitations to flight in turbulence. The energetic cost of flight may increase due to possible aerodynamic penalty on the efficiency of flapping wings or because of kinematic changes that the animals implement to stabilize their body position. Engels et al. [58] conducted a computational study of a tethered bumblebee model in homogeneous isotropic turbulence. It revealed that the efficiency of bumblebee wings is not sensitive to the background turbulence, as no significant changes in the statistical averages of the aerodynamic forces, moments, or aerodynamic power incurred, compared to steady inflow conditions. The instantaneous aerodynamic forces and torques, however, showed fluctuations becoming larger with the turbulence intensity, which can cause aerodynamic failure in free flight. Fisher et al. [63] used a mechanical flapping wing installed in a wind tunnel to explore the effect of turbulence across a range of flapping frequencies and turbulence intensities. The study showed that at lower flapping frequencies, the background turbulence dominates the near field of the wing and the lift fluctuation amplitude relative to its mean value is large. At higher flapping frequencies, flapping motion dominates and the lift fluctuates relatively less. This remarkable robustness of flapping wings as a lifting and propulsive device implies that the increased energetic cost of flapping flight in turbulence, if any, is related to the variation of body orientation that was found in the free-flight experiments.

As the body orientation changes, the vertical projection of the aerodynamic force becomes smaller. The aerodynamic force magnitude has to be consistently increased to maintain level flight. This is achieved by increasing the wingbeat frequency or amplitude or both [140, 152, 142, 40]. The increase comes at an added energetic cost, as the aerodynamic power required for flight grows roughly quadratically with the roll amplitude [99]. This cost can possibly be offset if the body posture is stabilized using asymmetric time variation of such kinematic parameters as the stroke amplitude, stroke plane angle, and feathering rotation angle. Yet, the energetic efficiency and limitations of these control strategies when applied to flight through turbulence need to be quantified. Apart from flapping flight, gliding is another mode of aerial locomotion likely to be affected by turbulence. Migrating birds, for instance, use thermal soaring to gain height, while minimizing the energy cost of propulsion. Thermals are ascending currents driven by the temperature gradient and inevitably producing strong turbulent fluctuations. Reddy et al. [154] explored the thermal soaring strategies optimal to different levels of atmospheric turbulence intensity using numerical models. The study combined numerical simulations of atmospheric flow with reinforcement learning methods. It showed that the optimal soaring strategies varied as the magnitude of turbulent fluctuations increased. In the regimes of weak turbulence, exploratory strategies were preferred, whereas in the regimes of strong turbulence, the optimal strategies were more risk-averse such as to reduce the probability of dramatic losses of height.

3.2.4 Biomimetics in Bioinspired Flight Systems

3.2.4.1 Overview on flapping micro–air vehicles

Flapping-wing propulsion, inspired by flying animals, possesses potential of efficient lift generation and gust mitigation capability and may provide an innovative solution to the dilemma of small autonomous MAVs. Biomimetic design systems associated with flapping MAVs can be classified into four key aspects: flapping mechanism, wing design, manufacturing, and systematic design. Two key issues in the biomimetic MAV design are flapping-wing mechanism and weight limitation. High power and high frequency are essential for the flapping-wing mechanism to create sufficient lift force. A light wing body is also a must, which, however, restricts the synthesis of the actuators, power sources, or materials, constraining the wing kinematic, frequencies, size, or available aerodynamics forces. Therefore, the main challenge in flapping MAV design and manufacturing, in particular, for insect of a smaller size, is how to realize wing kinematics within the crucially restricted mass [194].

Most flapping MAVs have been developed with a wingspan of 10^{-2} to 10^{0} m and a mass of 10^{-2} to 10^{0} kg. Electric motors and gear-crank linkage mechanisms are normally used to generate flapping-wing motions through converting the motor's rotational motion to the linkage's straight motion. For larger MAVs [90] , the payload can afford some avionics such as vision systems or autonomous control systems; and by processing the captured images with a ground station, autonomous obstacle avoidance can be realized. In these electric MAVs, the flapping motion is generated in terms of one degree of freedom (DOF), which can create feathering motion (AoA) owing to passive deformation of the flapping wing. Maneuvering flight is commonly achieved with a rudder or elevators of the tail wing as used in fixed-wing or rotary-wing aircraft. There are also MAVs capable of actively controlling feathering motion of the wing so as to realize maneuvering without rudders or elevators [194]. Despite the challenges for small-scale propulsion and miniaturization, researchers have successfully reproduced the bioinspired flight systems in several different ways, which may be classified into three prototypes of the four-winged (or X-wing) DelFly [45, 137] , the hummingbird-sized Nano-Hummingbird [90], and the insect-size Robobee [129].

The X-wing MAV has four flexible wings as two paired wings and uses the clap-and-fling mechanism achieved by the gear-crank linkage system [45, 137]. Aerodynamically, the X-wing MAVs benefit from three clap-and-fling cycles in a wingbeat, capable of augmenting aerodynamic force generation. One of the most successful bioinspired MAVs so far is the two-winged Nano-Hummingbird [90] with a wingspan of 16.5 cm and a weight of 19 g. All necessary components, including actuators, power source, flapping mechanism, and sensors for flight control, are put on board. Stable hovering and agile maneuverability are performed by active modulation of the flapping wings. The insect-size MAV, Robobee [129], presents one of the smallest flapping MAVs, with a wingspan of 30 mm, a weight of 80 mg, and a flapping frequency of 120 Hz. Robobee is equipped with high-power density, piezoelectric motors, which, however, restricts it to achieve real free flight. Manufactured by a novel methodology capable of rapidly prototyping articulated, flexure-based submillimeter mechanisms, the flyer can achieve a tethered but unconstrained stable hovering and basic controlled flight maneuvers.

Enhancement of the energy transformation from a motor to a mechanical flapping-wing system is also very challenging but remains yet poorly understood, which may significantly regulate the wing stroke dynamics and hence passively enhance versatile aerodynamic force generation. Harne et al. [75] investigated the dipteran wing motor in insects as a compression "bistable click" mechanism and found that a flight mechanism capable of adjusting motor axial support stiffness and compression characteristics may dramatically modulate the amplitude range and type of wing stroke dynamics achievable. Such flexible structures associated with motor and wing hinge may be of potential in enhancing the flexibility and robustness of the control autonomy system through integrating passive and active mechanisms interactively and complementarily. An overview of recent advances in biomimetic flapping-wing MAVs can be found, for example, in Ref. [123] with a specific focus on wing design and fabrication as well as sensing systems.

3.3 Unsteady Biomechanical Mechanisms in Swimming

Swimming organisms cover a much wider range in swimming speed and body size and possess more freedom in choosing propulsive kinematics, as well as morphological characteristics correspondingly.

Due to the complex nature of hydrodynamics and technical restriction, a qualitative understanding of swimming biofluid dynamics started with a few exquisite analytical models [116, 175] based on sufficient simplicity that the undulatory swimming is considered an undulating body moving in a steady flow, namely tethered swimming modeling. With introduction of the added mass concept [191], the unsteady swimming problem can be largely simplified, capable of being solved in a manner of quasi-steady estimation in the scope of potential flow theory [44]. Since the LEV has been found to be critical for lift generation in flapping flight [55, 56], unsteady hydrodynamic mechanisms in swimming have also become a main subject, which normally focuses on transient flow fields and their association with force production [47]. Meanwhile, with the remarkable developments in PIV [52, 135], high-resolution 3D tracking systems and kinematics measurements [185], computational fluid dynamics (e.g., [197, 111]), and swimming robotics (e.g., [150, 144]), quantitative analysis of highly unsteady hydrodynamics in swimming with high-resolved flow fields in space and time is now available.

Here we overview recent advances in the unsteady mechanisms in terms of unsteady hydrodynamics in steady swimming, hydrodynamics in C-start and maneuvering, and swimming hydrodynamics in unsteady environments.

3.3.1 Unsteady Hydrodynamics in Cyclic Swimming

In fish swimming, for both body and caudal fin (BCF) and median and paired fin (MPF) modes [160], steadily cyclic swimming via oscillation (flapping) or undulation is dynamically maintained by periodic wave motion of body or appendage segments, which is normally featured by unsteady hydrodynamics. Such unsteadiness has the potential to challenge the conclusion based on the quasi-steady models: Lau and Kelso [101] investigated the flow around a submerged fish-inspired hydrofoil and compared the hydrodynamic force estimated by the quasi-steady model to experimental force measurements. The experimental results show good agreement with the quasi-steady model when unsteady effects are small. However, when the parameter setting enters a zone where unsteady dynamic stall is expected to occur, the experimental results dramatically depart from the quasi-steady model as well as the scaling laws [101].

3.3.1.1 Flapping-fin hydrodynamics

Appendages such as pectoral fins of fish normally perform flapping motions as wings of insects and birds. Therefore, the AR, translating and rotating kinematics, and AoA are important parameters affecting the unsteady hydrodynamics in flapping fins [94]; fin oscillation is a means to generate and keep the vortices attached to the fin [9].

Active and passive deformations of oscillating pectoral fins are an active subject nowadays. At high AoAs, it is observed that pectoral fins of the bluegill sunfish (*Lepomis macrochirus*) could form edge vortices simultaneously on its upper and lower edges during part of the cycle [102, 9]. Such fin deformation can change the projected area in different planes and the directions of jets, which may benefit station-keeping performance in streams of low speeds [9]. The sunfish's pectoral fins produce high propulsive performance by employing a complex fin gait driven by active and passive fin deformations, and during pectoral fin abduction, the fin tip undergoes a complex, 3D flapping motion [50] that produces a strong and long-lasting, attached tip vortex. This tip vortex is associated with most of the thrust production during the abduction phase of the stroke [50]. In another delicate model built for mantas but implemented at an intermediate Re relatively lower than the factual Re [119, 62], kinematic analyses revealed that the fin movements are asymmetrical up-and-down strokes with both span- and chordwise waves interposed into the flapping motions [62]. These motions produce complex 3D flow patterns as well. Two sets of interconnected vortex rings are observed in the wake right behind each fin, which induce strong backward-flow jets and are responsible for the fin thrust generation [119]. The net thrust was produced from the distal half of the fins [62].

Fin flexibility can be beneficial to its performance, however, provided it is in proper spatial distribution. In the simulation by Shoele and Zhu [163], with a fully coupled FSI model, three fin models in labriform swimming morphology and kinematics were tested: (i) a flexible fin with a strengthened leading-edge ray, (ii) a flexible fin without a strengthened leading-edge ray, and (iii) a fin with rigid rays. Without leading-edge ray strengthening, the propulsion performance of the flexible fin is greatly compromised compared to the rigid-ray fin, suggesting that the stiffness of the ray at the leading edge plays a pivotal role in performance enhancement [163].

For the tail fin, normally driven by undulatory waves, we categorize it as an oscillating appendage here since it is usually considered stiff and thus its cross sections move synchronously. Beem et al. [15] investigated the influence of spanwise flow on the development and stabilization of LEVs on a foil (without rotational acceleration) and observed for both the LEV and the tip vortex that higher sweepback angles conduce to maintain a stronger connection between LEV and tip vortex [15]. Evidence of the vortex reattachment at the leading edge of a factual-shaped fish tail was obtained through high-resolution 3D simulations [21], and an attached LEV was observed at a small Strouhal number (St) of approximately 0.25 but vanished at St = 0.6. The LEV drastically alters the pressure distribution on the tail, while the flapping-fin kinematics is capable of stabilizing the LEV [21].

3.3.1.2 Undulatory swimming

In a classical view on undulatory swimming, the propulsive thrust is generated by pushing the surrounding water rearward, yet recent studies show that undulatory fish such as lampreys primarily pull themselves through the water via suction rather than pushing against the surrounding fluid [67, 68]. Effective generation in low-pressure regions are significant, while the unsteady mechanism has the potential to strengthen the low-pressure regions [111].

The 3D wake structure behind the fish has been investigated experimentally (e.g., [52, 135]) and computationally [22, 23]. Among various wake patterns observed, the general wake structure “single and double row vortices” is suggested to depend primarily on the

Strouhal number [22]. However, the tailing wake topology might have been overemphasized [21], because it would be difficult to quantify the interaction between near-field flow fields and surface stresses.

When the whole fish body participates in undulation, the boundary layer is destabilized as undulatory amplitude increases along the posterior body [5, 135]. As a result, a fish undergoing undulatory swimming will experience a larger friction drag compared to the same fish stretched straight (dead friction drag) in the flow [5]. Numerical studies suggest that the undulation along the posterior body is the main resource of such extra drag [111]. The wake vortex topology induced owing to such kinematics cannot be simply created by the tail (see the top view of the trailing edge of the fish), but is likely preformed and intensively interacts with the body (e.g., [135]). Patches of elevated vorticity form in a staggered pattern and travel posteriorly along with the body wave (Fig. 3.4b,c), until they are ultimately shed off at the tail (e.g., [135]). Unsteady mechanisms may takes part during the back and forth (left and right) motion of body segments and the generation and evolution of the elevated vorticity. Such an undulating body does not have a clear leading and trailing edge (Fig. 3.4a,b). Instead, the dorsal and ventral (i.e., upper and lower) edges of the undulatory body are equivalent. Using computational modeling, verified by PIV experiments, unsteady vortices around the upper and lower edges of the larval zebrafish at intermediate Reynolds number are identified [111] (Fig. 3.4c). Thrust contributions peak adjacent to the upper and lower edges of the fin fold where large low-pressure zones occur; thus propulsion in larval fish partly relies on unsteady high-intensity vortices along the upper and lower edges of the body [111] (see Fig. 3.4d).

Not limited to a typical elongated fish shape, in Rajiform locomotion, the stingray undulates its pectoral disc with complex 3D waves [18]. A horseshoe vortex was discovered to be present at the anterior margin of the stingray, creating a low-pressure region that enhances thrust for both fast and slow swimming speeds [25]. The LEV on the pectoral disk generates a low-pressure region in the fast-swimming stingray, enhancing thrust production in fast-swimming stingrays and improving the efficiency of swimming [25]. A similar mechanism is also observed in knifefish, which propel themselves by undulatory ribbon dorsal fins by generating a streamwise central jet, as well as the attached vortex rings [162].

In the undulating structure, the motions of cross sections are phase-shifted along the structure. Hence, the undulatory surface forms more complicated vortex structures than those found near oscillating surfaces: multiple pairs of vortices form in a staggered pattern and move relative to the body, owing to the body wave [111], that is, the undulatory form of the edge vortex.

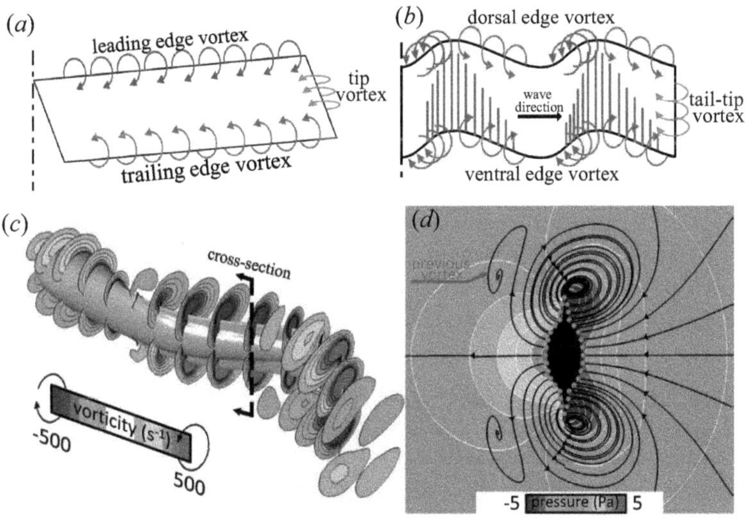

Figure 3.4 (a) Oscillating plate with attacking angle may form vortices respectively on the leading edge, the trailing edge, and the tip. However, (b) in an undulating plate, vortex structures in staggered pattern are formed on the upper (dorsal) and lower (ventral) edges. This undulation vortex topology is distinguished from that of the oscillation. (c) As a concrete example, in a larval fish swimming simulation by Li et al. [111], vertical cross sections of edge vortex structures are shown. (d) In-plane streamline and pressure fields on a typical cross section, position marked in (c), and the vertices occupying the fin edges coexist with the previously shed vortices. The low-pressure zones on the edges of the fin fold are related to the flow separation and reattachment [111]. (c, d) Republished with permission of *Journal of the Royal Society Interface*, from Ref. [111]; permission conveyed through Copyright Clearance Center, Inc.

3.3.1.3 Interaction between body and fins

While the body and fins generate and interact with vortex flows individually, hydrodynamic interactions exist between body and fins, and/or between fins, in a manner of vortex induction and capture. Tytell [179] proposed a 3D model of linked vortices induced by dorsal, anal, and caudal fins in sunfish swimming, revealing a complicated hydrodynamic interaction among the fins. Such interaction is suggested to be functional to angle the tail fin correctly to enhance thrust production [179]. The experimental study of trout swimming by Standen and Lauder [170] also showed that the flow was actively modified by dorsal and anal fins, resulting in complex flow patterns surrounding the caudal fin. The caudal fin does not encounter free-stream flow but rather moves through incident flow greatly altered by the action of dorsal and anal fins. A CFD study further confirmed that the vortex shedding from the upstream (dorsal) fin was indeed capable of increasing the thrust of the downstream (tail) fin significantly. However, this thrust augmentation was found to be sensitive to the phase relationship between the two flapping fins [4].

With respect to the interaction between pectoral fin and body, hydrodynamic interaction between the pectoral-fin-induced vortices and undulating fish body exists. Yu et al. [204] found a pair of vortices formed immediately behind the abducted pectoral fins of a swimming fish. The low-pressure suction forces owing to the shed pectoral fin vortices facilitate lateral movements of the fish body, decreasing the power consumption of the fish body [204]. However, comprehensive studies on the total energetic performance counting in both body and fin in such body–fin interaction are still needed.

3.3.2 Hydrodynamics in C-Start and Maneuvering

In nature, unsteady swimming is normal, in particular when swimmers perform a burst and/or complex maneuvers. Fish usually perform a C-start for escape. The C-start may be considered as a swimming mode of rapid acceleration and maneuvering. During C-start, fish initiate a swimming bout by bending their bodies into a C shape, while performing rotation in the first tail beat (referred as to stage 1), and then accelerate in the second tail beat (referred as to stage 2) [135, 180]. C-start hydrodynamics has been well studied both experimentally [135, 180] and computationally [108–110, 66]. During the escape response, fish can produce three distinct jets at the head, tail, and concave bend [180, 110] and further shed vortex pairs in the horizontal plane of the wake for each tail beat [135, 109]. Hydrodynamic performance of the C-start escape therefore relies on the capability of pronounced C-bent body configurations to trap and accelerate large volumes of fluid [66, 110]. Curvature modification along the body may essentially change the performance of the acceleration, rotation, and motion trajectory [66, 110]. As a unique and novel unsteady mechanism, when fish with an elongated body bend and take a U turn, in stage 2, it may even intercept their own wake shed in stage 1 and their trajectory become deflected by several degree. Wake interception causes no significant acceleration tangential to the trajectory of the fish and does not affect total power output [110].

With multiple fins, the C-start of the bluegill sunfish forms a wake in a more complex 3D structure than that of a simple elongated foil, with multiple connected vortex loops at the end of the C-start [24, 20]. In addition to the contribution of propulsion by the BCF, the dorsal and anal fins also provide considerable propulsive forces [180]; meanwhile, active control of the anal/dorsal fins can be related to retaining the stability of the sunfish against roll and pitch movements during the C-start and stabilizes the fish during C-start [19]. Unsteady escaping performance is also constrained by pre-existing motion and flow [6]: escapes executed during swimming were kinematically less variable than those made in still water, and it's worth noting that the performance of escapes further made up- and downstream differentiates.

During maneuvering, the body cross-section shape may also bring in an additional unsteady hydrodynamic effect. In contrast to the BCF fish swimmers, boxfish possess rigid, keeled exteriors. Though their body is incapable of bending, extraordinary maneuverability is observed—boxfish can even spin around with minimum turning radii near zero [186]. Behind the extraordinary maneuverability is the debate on the role of the unsteady hydrodynamics induced by the body cross-section shape. Bartol et al. observed, in an experimental study of boxfish, that the ventrolateral keels produce LEVs that may cause self-correcting forces, facilitating stableness in swimming [11, 12]. However, van Wassenbergh considers that tight swimming course stabilization is paradoxical, given the frequent and high-performance maneuvering of boxfish. Experimentally and computationally, they further identified the destabilizing moments due to turbulence consistently on the boxfish body and pointed out that such destabilizing moments might enhance maneuverability

[182]. While the occurrence of unsteady flow topology is confirmed in both studies, the unsteadiness effect remains to be further confirmed for boxfish.

3.3.3 Swimming Hydrodynamics in Unsteady Environments

3.3.3.1 Swimming in turbulence

In natural turbulence environments, fish need to adopt different kinematics strategies so as to stabilize their body attitudes at the least energy cost [76, 201]. The turbulence level matters a lot. When the turbulence is intensively broken down, fish tend to evade high intensity of turbulence [39]. Otherwise, as measured by Enders et al. [57] through respirometry experiments on juvenile Atlantic salmon (*Salmo salar*), the energetic cost for holding position in turbulence may increase several-folds. Turbulence may also decrease critical swimming speeds, while the size and spinning orientation eddies also make a difference [177]. In observations on labriform fish, the shiner surfperch, individuals in low-amplitude, unsteady flows performed as well as fish in steady flows. However, swimming costs in high-amplitude, unsteady flows were observed on average 25.3% higher than in steady flows and 14.2% higher than estimated values obtained from simulations based on the nonlinear relationship between swimming speed and oxygen consumption rate in steady flows [157]. On the contrary, if the turbulence is of high regularity (a well-known example is the von Kármán vortex street), and its structure is of a similar size comparable to the fish, the fish may be capable of exploiting the turbulence and reduce the energy cost (e.g., [115, 113]).

With respect to the interaction between fish body and the Kármán vortex street, Liao et al. [115, 114] found that trout swimming behind cylinders adopt a distinctive pattern of movement in order to remain stable. They termed it "Kármán gait" [115]. Trout in Kármán gait alter their body kinematics to synchronize with the shed vortices, with the tail-beat frequency distinctively reduced; matching the induced vortex shedding frequency from the cylinder [115, 114, 1]. Kármán-gaiting fish also exhibit substantial lateral translations and body rotations, while an inverse correlation between head angle and body angle was also observed [2]. Adopting a Kármán gait may substantially lead to a decrease in the cost of locomotion according to the measurement of tail-beat frequency and oxygen consumption [114, 173].

The Kármán gait may hardly require any active control but results from passive deformation. In the wake of a bluff cylinder, without expending any energy, a dead fish or a passively mounted high-AR foil may extract sufficient energy from the oncoming vortices to increase the thrust against drag [13]. Toming et al. [176] tested the hydrodynamics of artificial trout with passive deformation in the von Kármán vortex street, and the Kármán gait showed robustness, even though the body length of the fish-like object exceeded the wake wavelength [176].

Application of the Kármán gait in bioinspired swimming robots could be a promising method to exploit energy in turbulence formed by obstacle or collective swimming. Key technical problems such as controlling the algorithm and vortex street sensing are being resolved: swimming kinematics of the Kármán gait can be modeled in an analytical manner with convincing accuracy [2]. The biomimetic lateral line [3, 184] and pressure sensor array [84] have been reported to be capable of discriminating von Kármán vortex streets from uniform flows and determine the orientation and position of the platform with respect to the incoming flow and the center axis of the Kármán vortex street [3, 184].

3.3.3.2 Collective swimming

The thrust wake behind a self-propelled swimming fish is essentially different from the drag wake behind a bluff/streamlined obstacle. A vortex street can be formed in the thrust wake, although in the reverse vorticity direction, since the thrust generation procedure induces a momentum of water opposite to the swimming direction immediately behind the fish [192]. To the best of our knowledge, so far no specified swimming gait has been reported to exploit energy from such thrust wake. In a 2D formation, when downstream fish are located at the midway between two upstream fishes' wake, vortices on its two flanks may induce flow velocity and reduce the relative velocity of the oncoming flow of downstream fish and decrease its energy consumption. Such diamond-shape formation is considered as the basic 2D optimum structure in fish schools [192].

Furthermore, the two upstream fishes' vortex streets may combine into a structure between them similar to Kármán vortex streets (see, Fig. 4 in Ref. [113]), which might turn to an exploitable energy form for Kármán gait [114].

The diamond formation optimization seems quite conditional. To realize the diamond formation optimization, fish in school need to produce single-row vortex rings to ensure that vortices are in uniform rotational direction on one flank of the vortex street. According to the wake topology research, kinematics of schooling fish need to perform a lower Strouhal number in swimming to ensure single-row vortex rings [22]. The 3D wakes possess a complicated ring-shape spatial structure [22, 69], which is fundamentally different from 2D wake structures produced by a cylinder with infinite height in a 3D experiment. The diamond formation optimization seems sensitive to the co-planarity level of the upstream and downstream fish, as well as the undulation of the upstream fish. Furthermore, computational results demonstrate that the wakes of fish at realistic Reynolds numbers in an inertial regime will break down and have a low chance for constructive vortex interaction [43].

Another physical mechanism may be comparatively less conditional: as two swimming fish sufficiently approach, between the fish, the relative flow speed may reduce due to the channeling effect, and collective swimming may become more efficient [192]. As direct computational proof, Daghooghi and Borazjani [43] showed that a fish in a rectangular school swam faster with decreasing lateral distance, while consuming similar amounts of energy. The fish in the rectangular school with the smallest lateral distance (0.3 fish lengths) swims 20% faster than a solitary individual, while consuming similar amount of energy [43]. Experimental evidence shows that in a single pair of fish, fish tend to swim in a synchronized gait at higher speeds [8]. Swimming modes are dominated by out-phase (half tail beat cycle phase deference) and in-phase (no phase difference) configurations. Such observed synchronization may be explained from an energy efficiency perspective [8], while further accurate investigation on energy consumption due to phase difference in paired fish is expected.

Gains in efficiency may be obtained more easily than expected. Hemelrijk et al. used a CFD model to investigate swimming efficiency of model fish traveling solitarily and in schools at various interindividual distances in different configurations (diamond, rectangular, phalanx, and line). Results showed that these fish swim more efficiently in a school than alone, except in a dense phalanx [77].

A beneficial position in swimming in schools also seems be affected by the scale (amount of fish) of the fish school. A recent numerical study showed that when two fish swim in a tandem configuration, it seems more beneficial for the upstream fish, due to the wake-splitting effect of the downstream fish that causes an enhancement of pressure in the wake of the upstream fish. For the downstream fish, it gains an advantage of drafting conditionally and its performance deteriorates at Strouhal numbers greater than 0.40 [91]. In an experimental observation of a school of four labriform fish, pectoral fin beat frequencies were significantly higher for schooling individuals swimming in front of the school relative to those swimming in the back. Trailing fish were estimated to benefit from a 25.6% reduction in oxygen consumption, depending on correlations between swimming speeds and pectoral fin beat frequency and between swimming speeds and oxygen consumption of solitary fish. Leading individuals in the school were estimated to have higher oxygen consumption than solitary individuals swimming at the same speed, depending on their higher pectoral fin frequency [85]. A similar result is also observed in schools consisting of eight juvenile mullets *Liza aurata*. At faster speeds, fish with a lower maximal metabolic rate and aerobic scope swam near the rear of the school. These trailing fish required fewer tail beats to swim at the same speed as individuals at the front of the school, indicating that posterior positions provide hydrodynamic benefits that reduce swimming costs [92].

3.3.4 Biomimetics in Bioinspired Swimming Systems

Fish are natural swimming machines shaped through millions of years of evolution and competition. To enhance swimming performance such as maneuverability and energy efficiency in man-made systems, scientists and engineers attempt to find inspiration from natural fish and mimic their biological anatomy and behavior [148].

3.3.4.1 Fish robots

To our knowledge, the earliest reported fish robot for scientific usage is RoboTuna developed at the Massachusetts Institute of Technology (MIT), USA [10]. Over the following two decades until now, numerous brilliant designs of fish robots mimicking various swimming types of natural fish have been realized.

Thunniform and carangiform fish robots (e.g., [10, 60]) possess a rigid anterior part similar to tuna and carp. The rigid anterior part is designed in a streamlined shape to reduce drag, in which the control system, power supply, and motor are usually deployed [10, 131]. These fish robots rely on the oscillation of the tail fin. Smooth and soft oscillating motions, if realized, would be more beneficial (e.g., [158]).

Some swimming robots mimic anguilliform natural swimmers such as the eel or snake. An anguilliform robot performs body undulation with a large amplitude and short wavelength and thus consists of multiple elements and joints [41, 130, 205]. A few novel amphibious snake robots are even able to swim and crawl [41, 205].

Besides the BCF-mode fish robot, other biomimetic designs target MPF-mode fish. Fins are used as the primary propulsor; hence the robot can possess a rigid body without undulation, which makes room for components and parts and eases the burden for controlling the body gesture. The labriform fish robot [168, 174] adopts the typical fish's streamlined shape and uses paired pectoral fins for locomotion. A few Rajiform designs don't use the typical fish shape but adopt a flat disc shape, mimicking the manta or stingray (e.g., [190, 189]).

Other designs mimic fish using elongated anal (or dorsal) fins. Knife fish keep their body relatively straight while swimming [42, 118, 125]. The knife fish robot [42] can realize straightforward locomotion with a rigid hull. Through varying critical fin kinematic variables such as frequency, amplitude, and wavelength of sinusoidal traveling waves, an optimal operational region in the fin's kinematic can be found [42].

Comprehensively, a few fish robot designs use both fins and BCF undulation (e.g., [188]). The boxfish's structure and swimming gait are also mimicked. In the Micro-autonomous Robotic Ostraciiform

(MARCO), the ostracifform locomotion is implemented with a pair of 2-DOF pectoral fins and a 1-DOF tail fin, while the body shape of the robot is similar to a boxfish body [97].

3.3.4.2 Control

The fundamental issue of fish robot control is the generation of body/fin waves. The undulation along the body or elongated fin [41, 42] is more difficult to be realized than back-and-forth flapping motion. The 2D and 3D undulations require an even more complicated gait-planning algorithm to smoothen and optimize the undulation [125, 126, 189]. However, 3D flapping motion of the pectoral fin is far complex than back-and-forth flapping and also requires planning and optimization [174].

For example, Chowdhury et al. [35] translated the BCF mode carangiform swimming behavior of a biological fish to a robotic fish. The robotic fish model (kinematics and dynamics) is integrated with the Lighthill mathematical model framework to generate posterior body undulatory movements. A central pattern generator (CPG) controller consists of coupled networks capable of producing coordinated oscillatory patterns of rhythmic activity, while receiving simple adjustment signals [188]. In the amphibious snake robot [41], Crespi and Ijspeert used a locomotion controller based on the biological concept of CPGs; furthermore, the gaits are optimized online rather than as an off-line optimization process [41].

Fish robots are also desired to implement schooling and teamwork. A platform to test and verify algorithms and strategies for cooperation of multiple underwater mobile robots has been built [161]. Multi-agent robotic fish systems with efficient collective control and collaboration mechanisms are being deployed to collect marine information such as water temperature and pollution level [159].

3.3.4.3 Development for actuators

Servomotors are commonly used as actuators of fish robots [26, 79, 124], especially for flapping motion. Meanwhile, to achieve smooth and precise undulation, novel actuator types and materials are being developed.

Liao et al. [112] presented a robot fish actuated by a wire-driven flapping propulsor. The propulsor is composed of a flexible backbone and two sets of controlling wires. By controlling the two sets of wires, the propulsor can generate oscillatory motion (C-motion) and undulatory motion (S-motion) [112]. Hydraulic power has also been used to actuate a soft body. In the robot fish design by Katzschmann et al. [89], a closed-circuit drive system that used water as a transmission fluid was used to actuate the soft body. Circulation of water through internal body channels provided control over the fish's caudal fin propulsion and yaw motion [89]. A flexible magnetic actuator driven by an external rotating magnetic field could create several gaits such as the bending vibration, twisting vibration, and their combination [96]. Fin structures of the fish robot nowadays have become more sophisticated with individually moveable fin rays, which can be controlled to generate complex tail conformations and motion programs ms [60].

Using soft actuators made up of smart materials, fish robots may become motorless and gearless [158]. These novel, smart materials include shape memory alloys (SMAs) (e.g., [95]), electroactive polymers (EAP) [178], and ionic polymer metal composite (IPMC) film (e.g., [33]). Utilizing both active and passive deformation of the smart materials may also reduce the complexity of the control parameter space—in the tadpole micro-robot by Kim et al. [93], the motion was controlled by changing the frequency and duty ratio of the input voltage, and their experimental results showed that this technique can accurately control the steering and swimming speed of the robot.

Swimming robots can even be made by the integration of biological components with artificial devices (biomechatronics). Herr and Dennis designed, built, and characterized a swimming robot actuated by two explanted frog semitendinosus muscles and controlled by an embedded microcontroller [78]. Recently, by patterning dissociated rat cardiomyocytes on an elastomeric body enclosing a microfabricated gold skeleton, a tissue-engineered biohybrid ray was created [144]. In this artificial animal, optical stimulation induced sequential muscle activation via serpentine-patterned muscle circuits, leading to coordinated undulatory swimming phototactically following a light cue [144].

3.4 Conclusion and Perspectives

Flying and swimming animals are the most sophisticated and ultimate locomotors on the earth and present an alternative paradigm of biomimetic designs of MAVs and fish robots, which, however, normally brings highly unsteady biofluid dynamics challenges in achieving complex maneuvers under unsteady environments. In this chapter, we have given an overview of the recent advances in the scope of unsteady biofluid dynamics with a specific focus on several important topics of unsteady mechanisms in regard with flying and swimming. Here, we further highlight the challenges and several open questions for further research.

- *LEV mechanisms*: Among the identified multiple unsteady aerodynamic mechanisms that are capable of enhancing force production, the LEV mechanism is the most universal and robust one, which has been confirmed in both flapping and revolving wings. Recent progress further opens up new perspective topics such as theoretical estimation of the LEV-induced aerodynamic force, the reduced order modeling of the LEV dynamics, etc. An additional complication arises in forward flight when the uniform inflow velocity is superposed with the flapping velocity. The LEV stability in that case depends on the advance ratio. Swimming animals can propel and control themselves through both oscillation fins and an undulating body. Flapping fins can create LEVs to improve performance; an undulating body can produce dorsal- and ventral-edge vortices in a staggered pattern. The edge vortex topology along the undulatory body shares some common feature of the LEV with that in flapping and revolving wings in a low-Reynolds-number regime. However, studies on the LEV-based mechanism in swimming are still very limited, and there is obviously large room to explore the Reynolds number effect of LEVs over a broad range of body sizes and swimming modes.
- *Passive mechanisms in flexible structures*: Flexibility in appendages can enhance the performance of locomotion by adjusting shape or kinematics adaptively. Most of the studies on the performance of flexible structures so far largely rely

on the simplified, or unrealistic, model of the wings, fins, or wing hinges due to the lack of a complete structural model. Accurate modeling faces great difficulties because of diverse, non-uniform, multiscale structures. From the biomimetic viewpoint, however, it is important to work out how flexible structures are designed and interact with unsteady fluid dynamics, because the adaptive adjustment can greatly simplify the biomimetic locomotor mechanism if properly designed. The main topic on the function of the flexible mechanism in biofluid dynamics has been its capability of force generation and power economy, while the flexible structure may also enhance its robustness by responding adaptively to environmental unsteadiness and randomness. The interaction between the integrated flexible model and unsteady, random flow obviously requires more investigation, not only for unveiling the novel mechanisms of bioinspired mechanical systems in flying and swimming, but also for abstracting biomimetic design principles, although such studies are very few, particularly in the field of animal flight.

- *Turbulence rejection and adaptation*: Flying and swimming animals employ various mechanisms to cope with turbulence, corresponding to the intensity, scale, and regularity level of the turbulence. For flying animals, stabilization against turbulence is vital; sophisticated strategies, including positive and passive adjustments on wingbeating kinematics, are observed in experiments. Applying such adjustments will increase the flying power consumption, which is also limited by physical conditions of the flyer. For swimmers, high intensity of turbulence can also increase the swimming power consumption. In contrast, if turbulence is of high regularity, such as the von Kármán vortex street, fish may adopt an effective strategy/gait to exploit energy from the turbulence and reduce energy cost. Turbulence rejection and adaptation are a challenging subject in regard with flying and swimming as an extreme of the unsteady biofluid dynamics, in which much remains unclear yet. Uncovering their novel mechanisms is therefore extremely important for bioinspired robotic designs in terms of achieving robustness and survival ability of flying and swimming in nature environments.

- *Biomimetic robotics*: Bioinspired flight and swimming systems as an integrated system [46] offer an alternative paradigm for micro-aerial and underwater vehicles when scaled down to insect, bird, and fish size, which, however, normally brings low-speed fluid dynamics and control challenges in achieving autonomous flight and swimming.
- Despite the challenges for small-scale propulsion and miniaturization, researchers have successfully reproduced bioinspired flapping-wing MAVs at the insect and bird scale. It is foreseen that rapid developments in biomimetic wing design and MEMS-based fabrication methods, as well as insect-inspired inertial and visual sensors along with bioinspired systematic design strategies, will bring more innovative designs. Conventional fixed-wing and rotary aircraft designs are normally performed with rigid wings and bodies designed separately and with aerodynamic theories and flight control laws developed separately. Bioinspired flight or swimming systems are, however, an integration of different flexible structures at different levels of body, wing, fin, musculoskeleton, sensors, and motors. We still know less about how these flexible structures work interactively and complementarily with active mechanisms through a closed loop of the inner working system and external mechanical system in order to achieve systematically efficient stabilization and control in flapping flight and swimming. Unveiling the passive and active mechanisms in concert with flexible structures in integrated bioinspired flight systems may offer a sophisticated solution to the trade-off in biological flights. This will further bring breakthroughs in integrating fluid dynamics and control strategies and hence a paradigm shift in biomimetic systematic design and miniaturization for small-scale autonomous unmanned aerial or underwater systems with capabilities to achieve controlled flight or swimming autonomy and to perform a variety of complex maneuvers.

Acknowledgments

HL and TN were partly supported by the Grant-in-Aid for Scientific Research on Innovative Areas No. 24120007, JSPS, and Tough

Robotics Challenge - Impulsing Paradigm Change through Disruptive Technologies Program, JST. TN was partly supported by JSPS KAKENHI Grant No. JP17K17638. GL was partly supported by JSPS KAKENHI Grant No. JP17K17641. DK acknowledges the financial support from the JSPS Postdoctoral Fellowship.

References

1. Akanyeti, O. and Liao, J. C. (2013a). The effect of flow speed and body size on kármán gait kinematics in rainbow trout, *J. Exp. Biol.*, **216**(18), pp. 3442–3449.
2. Akanyeti, O. and Liao, J. C. (2013b). A kinematic model of kármán gaiting in rainbow trout, *J. Exp. Biol.*, **216**(24), pp. 4666–4677.
3. Akanyeti, O., Venturelli, R., Visentin, F., Chambers, L., Megill, W. M. and Fiorini, P. (2011). What information do kármán streets offer to flow sensing? *Bioinspir. Biomim.*, **6**(3), p. 036001.
4. Akhtar, I., Mittal, R., Lauder, G. V. and Drucker, E. (2007). Hydrodynamics of a biologically inspired tandem flapping foil configuration, *Theor. Comp. Fluid Dyn.*, **21**(3), pp. 155–170.
5. Anderson, E. J., McGillis, W. R. and Grosenbaugh, M. A. (2001). The boundary layer of swimming fish, *J. Exp. Biol.*, **204**(1), pp. 81–102.
6. Anwar, S. B., Cathcart, K., Darakananda, K., Gaing, A. N., Shin, S. Y., Vronay, X., Wright, D. N. and Ellerby, D. J. (2016). The effects of steady swimming on fish escape performance, *J. Comp. Physiol. A*, **202**(6), pp. 425–433.
7. Appel, E. and Gorb, S. N. (2011). Resilin-bearing wing vein joints in the dragonfly epiophlebia superstes, *Bioinspir. Biomim.*, **6**, p. 046006.
8. Ashraf, I., Godoy-Diana, R., Halloy, J., Collignon, B. and Thiria, B. (2016). Synchronization and collective swimming patterns in fish (Hemigrammus bleheri), *J. R. Soc. Interface*, **13**(123), p. 20160734.
9. Bandyopadhyay, P. R., Beal, D. N. and Menozzi, A. (2007). Biorobotic insights into how animals swim, *J. Exp. Biol.*, 211(2), pp. 206–214.
10. Barrett, D., Grosenbaugh, M. and Triantafyllou, M. (1996). The optimal control of a flexible hull robotic undersea vehicle propelled by an oscillating foil, in *Proceedings of Symposium on Autonomous Underwater Vehicle Technology*, pp. 1–9.
11. Bartol, I. K., Gharib, M., Webb, P. W., Weihs, D. and Gordon, M. S. (2005). Body-induced vortical flows: a common mechanism for self-corrective trimming control in boxfishes, *J. Exp. Biol.*, **208**(2), pp. 327–344.

12. Bartol, I. K., Gordon, M. S., Webb, P., Weihs, D. and Gharib, M. (2008). Evidence of self-correcting spiral flows in swimming boxfishes, *Bioinspir. Biomim.*, **3**(1), p. 014001.

13. Beal, D. N., Hover, F. S., Triantafyllou, M. S., Liao, J. C. and Lauder, G. V. (2006). Passive propulsion in vortex wakes, *J. Fluid. Mech.*, **549**, pp. 385–402.

14. Beatus, T. and Cohen, I. (2015). Wing-pitch modulation in maneuvering fruit flies is explained by an interplay between aerodynamics and a torsional spring, *Phys. Rev. E*, **92**, p. 022712.

15. Beem, H. R., Rival, D. E. and Triantafyllou, M. S. (2012). On the stabilization of leading-edge vortices with spanwise flow, *Exp. Fluids*, **52**(2), pp. 511–517.

16. Bergou, A. J., Ristroph, L., Guckenheimer, J., Cohen, I. and Wang, Z. (2010). Fruit flies modulate passive wing pitching to generate in-flight turns, *Phys. Rev. Lett.*, **104**, p. 148101.

17. Bergou, A. J., Xu, S. and Wang, Z. J. (2007). Passive wing pitch reversal in insect flight, *J. Fluid Mech.*, **591**, pp. 321–337.

18. Blevins, E. L. and Lauder, G. V. (2012). Rajiform locomotion: three-dimensional kinematics of the pectoral fin surface during swimming in the freshwater stingray potamotrygon orbignyi, *J. Exp. Biol.*, **215**(18), pp. 3231–3241.

19. Borazjani, I. (2013). The functional role of caudal and anal/dorsal fins during the c-start of a bluegill sunfish, *J. Exp. Biol.*, **216**(9), pp. 1658–1669.

20. Borazjani, I. (2015). Simulations of unsteady aquatic locomotion: from unsteadiness in straight-line swimming to fast-starts, *Integr. Comp. Biol.*, **55**(4), p. 740.

21. Borazjani, I. and Daghooghi, M. (2013). The fish tail motion forms an attached leading edge vortex, *Proc. R. Soc. London, Ser. B*, **280**(1756).

22. Borazjani, I. and Sotiropoulos, F. (2008). Numerical investigation of the hydrodynamics of carangiform swimming in the transitional and inertial flow regimes, *J. Exp. Biol.*, **211**(10), pp. 1541–1558.

23. Borazjani, I. and Sotiropoulos, F. (2009). On the role of form and kinematics on the hydrodynamics of self-propelled body/caudal fin swimming, *J. Exp. Biol.*, **213**(1), pp. 89–107.

24. Borazjani, I., Sotiropoulos, F., Tytell, E. D. and Lauder, G. V. (2012). Hydrodynamics of the bluegill sunfish c-start escape response: three-dimensional simulations and comparison with experimental data, *J. Exp. Biol.*, **215**(4), pp. 671–684.

25. Bottom II, R. G., Borazjani, I., Blevins, E. L. and Lauder, G. V. (2016). Hydrodynamics of swimming in stingrays: numerical simulations and the role of the leading-edge vortex, *J. Fluid. Mech.*, **788**, pp. 407–443.

26. Cai, Y., Bi, S. and Zheng, L. (2012). Design optimization of a bionic fish with multi-joint fin rays, *Adv. Rob.*, **26**(1–2), pp. 177–196.

27. Carr, Z. R., DeVoria, A. C. and Ringuette, M. J. (2015). Aspect-ratio effects on rotating wings: circulation and forces, *J. Fluid. Mech.*, **767**, pp. 497–525.

28. Carruthers, A. C., Thomas, A. L. R. and Taylor, G. K. (2007). Automatic aeroelastic devices in the wings of a steppe eagle aquila nipalensis, *J. Exp. Biol.*, **210**, pp. 4136–4149.

29. Chapman, J. W. and Drake, V. A. (2010). Insect migration, in *Encyclopedia of Animal Behavior*, Vol. 2 (Elsevier), ISBN 0066-4170, pp. 161–166.

30. Chen, D., Kolomenskiy, D. and Liu, H. (2017). Closed-form solution for the edge vortex of a revolving plate, *J. Fluid. Mech.*, To be published.

31. Chen, J.-S., Chen, J.-Y. and Chou, Y.-F. (2008). On the natural frequencies and mode shapes of dragonfly wings, *J. Sound Vibr.*, **313**, pp. 643–654.

32. Chen, Y., Gravish, N., Desbiens, A. L., Malka, R. and Wood, R. J. (2016). Experimental and computational studies of the aerodynamic performance of a flapping and passively rotating insect wing, *J. Fluid Mech.*, **791**, pp. 1–33.

33. Chen, Z., Shatara, S. and Tan, X. (2010). Modeling of biomimetic robotic fish propelled by an ionic polymer-metal composite caudal fin, *IEEE/ASME Trans. Mechatron.*, **15**(3), pp. 448–459.

34. Cheng, B., Sane, S. P., Barbera, G., Troolin, D. R., Strand, T. and Deng, X. (2013). Three-dimensional flow visualization and vorticity dynamics in revolving wings, *Exp. Fluids*, **54**(1), p. 1423.

35. Chowdhury, A., Kumar, V., Prasad, B., Kumar, R. and Panda, S. (2014). Kinematic study and implementation of a bio-inspired robotic fish underwater vehicle in a lighthill mathematical framework, *Rob. Biomimetics*, **1**(1), p. 15.

36. Combes, S. A. and Daniel, T. L. (2003a). Flexural stiffness in insect wings I. Scaling and the influence of wing venation, *J. Exp. Biol.*, **206**(17), pp. 2979–2987.

37. Combes, S. A. and Daniel, T. L. (2003b). Into thin air: contributions of aerodynamic and inertial-elastic forces to wing bending in the hawkmoth Manduca sexta, *J. Exp. Biol.*, **206**(17), pp. 2999–3006.

38. Combes, S. A. and Dudley, R. (2009). Turbulence-driven instabilities limit insect flight performance, *Proc. Natl. Acad. Sci. U. S. A.*, **106**(22), pp. 9105–9108.

39. Cotel, A. J., Webb, P. W. and Tritico, H. (2006). Do brown trout choose locations with reduced turbulence? *Trans. Am. Fish. Soc.*, **135**(3), pp. 610–619.

40. Crall, J. D., Chang, J. J., Oppenheimer, R. L. and Combes, S. A. (2017). Foraging in an unsteady world: bumblebee flight performance in field-realistic turbulence, *Interface Focus*, **7**(1), p. 20160086.

41. Crespi, A. and Ijspeert, A. J. (2008). Online optimization of swimming and crawling in an amphibious snake robot, *IEEE Trans. Rob.*, **24**(1), pp. 75–87.

42. Curet, O. M., Patankar, N. A., Lauder, G. V. and MacIver, M. A. (2011). Mechanical properties of a bio-inspired robotic knifefish with an undulatory propulsor, *Bioinspir. Biomim.*, **6**(2), p. 026004.

43. Daghooghi, M. and Borazjani, I. (2015). The hydrodynamic advantages of synchronized swimming in a rectangular pattern, *Bioinspir. Biomim.*, **10**(5), p. 056018.

44. Daniel, T. L. (1984). Unsteady aspects of aquatic locomotion, *Am. Zool.*, **24**(1), pp. 121–134.

45. de Croon, G., Groen, M., Wagter, C., Remes, B., Ruijsink, R. and van Oudheusden, B. (2012). Design, aerodynamics and autonomy of the delfly, *Bioinspir. Biomim.*, p. 025003.

46. Dickinson, M. (2006). Insect flight, *Curr. Biol.*, **16**, pp. R309–R314.

47. Dickinson, M. H. (1996). Unsteady mechanisms of force generation in aquatic and aerial locomotion, *Am. Zool.*, **36**(6), pp. 537–554.

48. Dickinson, M. H., Farley, C. T., Full, R. J., Koehl, M. A. R., Kram, R. and Lehman, S. (2000). How animals move: an integrative view, *Science*, **288**, pp. 100–106.

49. Dickinson, M. H. and Lighton, J. R. B. (1995). Muscle efficiency and elastic storage in the flight motor of drosophila, *Science*, **268**, pp. 87–90.

50. Dong, H., Bozkurttas, M., Mittal, R., Madden, P. and Lauder, G. V. (2010). Computational modelling and analysis of the hydrodynamics of a highly deformable fish pectoral fin, *J. Fluid. Mech.*, **645**, pp. 345–373.

51. Donoughe, S., Crall, J. D., Merz, R. A. and Combes, S. A. (2011). Resilin in dragonfly and damselfly wings and its implications for wing flexibility, *J. Morphol.*, **272**, pp. 1409–1421.

52. Drucker, E. G. and Lauder, G. V. (1999). Locomotor forces on a swimming fish: three-dimensional vortex wake dynamics quantified using digital particle image velocimetry, *J. Exp. Biol.*, **202**(18), pp. 2393–2412.

53. Du, G. and Sun, M. (2010). Effects of wing deformation on aerodynamic forces in hovering hoverflies, *J. Exp. Biol.*, **213**, pp. 2273–2283.

54. Eldredge, J. D., Toomey, J. and Medina, A. (2010). On the roles of chord-wise flexibility in a flapping wing with hovering kinematics, *J. Fluid Mech.*, **659**, pp. 94–115.

55. Ellington, C. P. (1995). Unsteady aerodynamics of insect flight, *Symp. Soc. Exp. Biol.*, **49**, p. 109129.

56. Ellington, C. P., van den Berg, C., Willmott, A. P. and Thomas, A. L. R. (1996). Leading-edge vortices in insect flight, *Nature*, **384**(6610), pp. 626–630.

57. Enders, E. C., Boisclair, D. and Roy, A. G. (2003). The effect of turbulence on the cost of swimming for juvenile atlantic salmon (salmo salar), *Can. J. Fish. Aquat. Sci.*, **60**(9), pp. 1149–1160.

58. Engels, T., Kolomenskiy, D., Schneider, K., Lehmann, F.-O. and Sesterhenn, J. (2016). Bumblebee flight in heavy turbulence, *Phys. Rev. Lett.*, **116**(2), p. 028103.

59. Ennos, A. R. (1988). The inertial cause of wing rotation in diptera, *J. Exp. Biol.*, **140**, pp. 161–169.

60. Esposito, C. J., Tangorra, J. L., Flammang, B. and Lauder, G. (2011). A robotic fish caudal fin: effects of stiffness and motor program on locomotor performance, *J. Exp. Biol.*, **215**(1), pp. 56–67.

61. Finnigan, J. (2000). Turbulence in plant canopies, *Annu. Rev. Fluid Mech.*, **32**(1), pp. 519–571.

62. Fish, F. E., Schreiber, C. M., Moored, K. W., Liu, G., Dong, H. and Bart-Smith, H. (2016). Hydrodynamic performance of aquatic flapping: efficiency of underwater flight in the manta, *Aerospace*, **3**(3).

63. Fisher, A., Ravi, S., Watkins, S., Watmuff, J., Wang, C., Liu, H. and Petersen, P. (2016). The gust-mitigating potential of flaping wings, *Bioinspir. Biomim.*, **11**(4), p. 046010.

64. Garmann, D. J. and Visbal, M. R. (2014). Dynamics of revolving wings for various aspect ratios, *J. Fluid. Mech.*, **748**, pp. 932–956.

65. Garmann, D. J., Visbal, M. R. and Orkwis, P. D. (2013). Three-dimensional flow structure and aerodynamic loading on a revolving wing, *Phys. Fluids*, **25**(3), p. 034101.

66. Gazzola, M., Rees, W. M. V. and Koumoutsakos, P. (2012). C-start: optimal start of larval fish, *J. Fluid. Mech.*, **698**, pp. 5–18.

67. Gemmell, B. J., Colin, S. P., Costello, J. H. and Dabiri, J. O. (2015). Suction-based propulsion as a basis for efficient animal swimming, *Nat. Commun.*, **6**, p. 8790.

68. Gemmell, B. J., Fogerson, S. M., Costello, J. H., Morgan, J. R., Dabiri, J. O. and Colin, S. P. (2016). How the bending kinematics of swimming lampreys build negative pressure fields for suction thrust, *J. Exp. Biol.*, **219**(24), pp. 3884–3895.

69. Green, M. A., Rowley, C. W. and Smits, A. J. (2011). The unsteady three-dimensional wake produced by a trapezoidal pitching panel, *J. Fluid. Mech.*, **685**, pp. 117–145.

70. Ha, N. S., Truong, Q. T., Goo, N. S. and Park, H. C. (2013). Relationship between wingbeat frequency and resonant frequency of the wing in insects, *Bioinspir. Biomim.*, **8**, p. 046008.

71. Haas, F., Gorb, S. and Blickhan, R. (2000). The function of resilin in beetle wings, *Proc. R. Soc. London, Ser. B*, **267**, pp. 1375–1381.

72. Harbig, R. R., Sheridan, J. and Thompson, M. C. (2013a). Relationship between aerodynamic forces, flow structures and wing camber for rotating insect wing planforms, *J. Fluid. Mech.*, **730**, pp. 52–75.

73. Harbig, R. R., Sheridan, J. and Thompson, M. C. (2013b). Reynolds number and aspect ratio effects on the leading-edge vortex for rotating insect wing planforms, *J. Fluid. Mech.*, **717**, pp. 166–192.

74. Harbig, R. R., Sheridan, J. and Thompson, M. C. (2014). The role of advance ratio and aspect ratio in determining leading-edge vortex stability for flapping flight, *J. Fluid. Mech.*, **751**, pp. 71–105.

75. Harne, R. and Wang, K. (2015). Dipteran wing motor-inspired flapping flight versatility and effectiveness enhancement, *J. R. Soc. Interface*, **12**, p. 20141367.

76. Heatwole, S. J. and Fulton, C. J. (2013). Behavioural flexibility in reef fishes responding to a rapidly changing wave environment, *Mar. Biol.*, **160**(3), pp. 677–689.

77. Hemelrijk, C. K., Reid, D. A. P., Hildenbrandt, H. and Padding, J. T. (2015). The increased efficiency of fish swimming in a school, *Fish Fish.*, **16**(3), pp. 511–521.

78. Herr, H. and Dennis, R. G. (2004). A swimming robot actuated by living muscle tissue, *J. NeuroEng. Rehabil.*, **1**(1), p. 6.

79. Hu, Y., Wang, L., Yu, J., Huo, J. and Jia, Y. (2008). Development and control of dolphin-like underwater vehicle, in *2008 American Control Conference*, pp. 2858–2863.
80. Huhn, F., van Rees, W., Gazzola, M., Rossinelli, D., Haller, G. and Koumoutsakos, P. (2015). Quantitative flow analysis of swimming dynamics with coherent lagrangian vortices, *Chaos*, **25**, p. 087405.
81. Ishihara, D. and Horie, T. (2017). Passive mechanism of pitch recoil in flapping insect wings, *Bioinspir. Biomim.*, **12**, p. 016008.
82. Ishihara, D., Horie, T. and Denda, M. (2009a). A two-dimensional computational study on the fluid-structure interaction cause of wing pitch changes in dipteran flapping flight, *J. Exp. Biol.*, **212**, pp. 1–10.
83. Ishihara, D., Yamashita, Y., Horie, T., Yoshida, S. and Niho, T. (2009b). Passive maintenance of high angle of attack and its lift generation during flapping translation in crane fly wing, *J. Exp. Biol.*, **212**, pp. 3882–3891.
84. Ježov, J., Akanyeti, O., Chambers, L. D. and Kruusmaa, M. (2012). Sensing oscillations in unsteady flow for better robotic swimming efficiency, in *2012 IEEE International Conference on Systems, Man, and Cybernetics (SMC)*, pp. 91–96.
85. Johansen, J. L., Vaknin, R., Steffensen, J. F. and Domenici, P. (2010). Kinematics and energetic benefits of schooling in the labriform fish, striped surfperch embiotoca lateralis, *Mar. Ecol. Prog. Ser.*, **420**, pp. 221–229.
86. Jones, A. R., Medina, A., Spooner, H. and Mulleners, K. (2016). Characterizing a burst leading-edge vortex on a rotating flat plate wing, *Exp. Fluids*, **57**(4), p. 52.
87. Jongerius, S. R. and Lentink, D. (2010). Structural analysis of a dragonfly wing, *Exp. Mech.*, **50**, pp. 1323–1334.
88. Kang, C.-K., Aono, H., Cesnik, C. E. S. and Shyy, W. (2011). effects of flexibility on the aerodynamic performance of flapping wings, *J. Fluid Mech.*, **689**, pp. 32–74.
89. Katzschmann, R. K., Marchese, A. D. and Rus, D. (2016). *Hydraulic Autonomous Soft Robotic Fish for 3D Swimming* (Springer International Publishing, Cham), ISBN 978-3-319-23778-7, pp. 405–420.
90. Keennon, M., Klingebiel, K., Won, H. and Andriukov, A. (2012). Development of the nano hummingbird: a tailless flapping wing micro air vehicle, *Aerospace Sciences Meeting including the New Horizons Forum and Aerospace Exposition*, AIAA 20112–20588.

91. Khalid, M. S. U., Akhtar, I. and Dong, H. (2016). Hydrodynamics of a tandem fish school with asynchronous undulation of individuals, *J. Fluids Struct.*, **66**, pp. 19–35.

92. Killen, S. S., Marras, S., Steffensen, J. F. and McKenzie, D. J. (2011). Aerobic capacity influences the spatial position of individuals within fish schools, *Proc. R. Soc. London, Ser. B*, **279**(1727), pp. 357–364.

93. Kim, B., Kim, D.-H., Jung, J. and Park, J.-O. (2005). A biomimetic undulatory tadpole robot using ionic polymer-metal composite actuators, *Smart Mater. Struct.*, **14**(6), p. 1579.

94. Kim, D. and Gharib, M. (2010). Experimental study of three-dimensional vortex structures in translating and rotating plates, *Exp. Fluids*, **49**(1), pp. 329–339.

95. Kim, H.-J., Song, S.-H. and Ahn, S.-H. (2013). A turtle-like swimming robot using a smart soft composite (ssc) structure, *Smart Mater. Struct.*, **22**(1), p. 014007.

96. Kim, S. H., Shin, K., Hashi, S. and Ishiyama, K. (2012). Magnetic fish-robot based on multi-motion control of a flexible magnetic actuator, *Bioinspir. Biomim.*, **7**(3), p. 036007.

97. Kodati, P., Hinkle, J., Winn, A. and Deng, X. (2008). Microautonomous robotic ostraciiform (marco): hydrodynamics, design, and fabrication, *IEEE Trans. Rob.*, **24**(1), pp. 105–117.

98. Kolomenskiy, D., Elimelech, Y. and Schneider, K. (2014). Leading-edge vortex shedding from rotating wings, *Fluid Dyn. Res.*, **46**, p. 031421.

99. Kolomenskiy, D., Ravi, S., Takabayashi, T., Ikeda, T., K., U., Engels, T., Fisher, A., Tanaka, H., Schneider, K., Sesterhenn, J. and Liu, H. (2017). Added costs of insect-scale flapping flight in unsteady airflows, *R. Soc. Open Sci.*, Manuscript in preparation.

100. Kruyt, J. W., van Heijst, G. F., Altshuler, D. L. and Lentink, D. (2015). Power reduction and the radial limit of stall delay in revolving wings of different aspect ratio, *J. R. Soc. Interface*, **12**(105), p. 20150051.

101. Lau, T. C. W. and Kelso, R. M. (2016). A scaling law for thrust generating unsteady hydrofoils, *J. Fluids Struct.*, **65**, pp. 455–471.

102. Lauder, G. V., Madden, P. G. A., Mittal, R., Dong, H. and Bozkurttas, M. (2006). Locomotion with flexible propulsors: I. Experimental analysis of pectoral fin swimming in sunfish, *Bioinspir. Biomim.*, **1**(4), p. S25.

103. Le, T. Q., Truong, T. V., Park, S. H., Truong, T. Q., Ko, J. H., Park, H. C. and Byun, D. (2013). Improvement of the aerodynamic performance by wing flexibility and elytra -hind wing interaction of a beetle during forward flight, *J. R. Soc. Interface*, **10**, p. 20130312.

104. Lehmann, F.-O., Gorb, S., Nasir, N. and Schützner, P. (2011). Elastic deformation and energy loss of flapping fly wings, *J. Exp. Biol.*, **214**, pp. 2949–2961.

105. Lentink, D. and Dickinson, M. H. (2009a). Biofluiddynamic scaling of flapping, spinning and translating fins and wings, *J. Exp. Biol.*, **212**(16), pp. 2691–2704.

106. Lentink, D. and Dickinson, M. H. (2009b). Rotational accelerations stabilize leading edge vortices on revolving fly wings, *J. Exp. Biol.*, **212**, pp. 2705–2719.

107. Lepora, N. F., Verschure, P. and Prescott, T. J. (2013). The state of the art in biomimetics, *Bioinspir. Biomim.*, **8**(1), p. 013001.

108. Li, G., Liu, H., Mller, U. K. and van Leeuwen, J. L. (2011). Swimming hydrodynamics and maneuverability in c-start of zebrafish larvae: an integrated computational study, *ASME-JSME-KSME 2011 Joint Fluids Engineering Conference: Volume 1, Symposia Parts A, B, C, and D*, pp. 2059–2066.

109. Li, G., Müller, U. K., van Leeuwen, J. L. and Liu, H. (2012). Body dynamics and hydrodynamics of swimming fish larvae: a computational study, *J. Exp. Biol.*, **215**(22), pp. 4015–4033.

110. Li, G., Müller, U. K., van Leeuwen, J. L. and Liu, H. (2014). Escape trajectories are deflected when fish larvae intercept their own c-start wake, *J. R. Soc. Interface*, **11**(101).

111. Li, G., Müller, U. K., van Leeuwen, J. L. and Liu, H. (2016). Fish larvae exploit edge vortices along their dorsal and ventral fin folds to propel themselves, *J. R. Soc. Interface*, **13**(116).

112. Liao, B., Li, Z. and Du, R. (2014). Robot fish with a novel biomimetic wire-driven flapping propulsor, *Adv. Rob.*, **28**(5), pp. 339–349.

113. Liao, J. C. (2007). A review of fish swimming mechanics and behavior in altered flows, *Philos. Trans. R. Soc. London, Ser. B*, **362**(1487), pp. 1973–1993.

114. Liao, J. C., Beal, D. N., Lauder, G. V. and Triantafyllou, M. S. (2003a). Fish exploiting vortices decrease muscle activity, *Science*, **302**(5650), pp. 1566–1569.

115. Liao, J. C., Beal, D. N., Lauder, G. V. and Triantafyllou, M. S. (2003b). The kármán gait: novel body kinematics of rainbow trout swimming in a vortex street, *J. Exp. Biol.*, **206**(6), pp. 1059–1073.

116. Lighthill, M. J. (1971). Large-amplitude elongated-body theory of fish locomotion, *Proc. R. Soc. London, Ser. B*, **179**(1055), pp. 125–138.

117. Limacher, E., Morton, C. and Wood, D. (2016). On the trajectory of leading-edge vortices under the influence of coriolis acceleration, *J. Fluid. Mech.*, **800**, p. R1.

118. Liu, F., Lee, K. M. and Yang, C. J. (2012). Hydrodynamics of an undulating fin for a wave-like locomotion system design, *IEEE/ASME Trans. Mechatron.*, **17**(3), pp. 554–562.

119. Liu, G., Ren, Y., Zhu, J., Bart-Smith, H. and Dong, H. (2015). Thrust producing mechanisms in ray-inspired underwater vehicle propulsion, *Theor. Appl. Mech. Lett.*, **5**(1), pp. 54–57.

120. Liu, H. (2009). Integrated modeling of insect flight: from morphology, kinematics to aerodynamics, *J. Comput. Phys.*, **228**, pp. 439–459.

121. Liu, H. and Aono, H. (2009). Size effects on insect hovering aerodynamics: an integrated computational study, *Bioinspir. Biomim.*, **4**(1), p. 015002.

122. Liu, H., Ellington, C. P., Kawachi, K., van den Berg, C. and Willmott, A. P. (1998). A computational fluid dynamic study of hawkmoth hovering, *J. Exp. Biol.*, **201**(4), pp. 461–477.

123. Liu, H., Ravi, S., Kolomenskiy, D. and Tanaka, H. (2016). Biomechanics and biomimetics in insect-inspired flight systems, *Philos. Trans. R. Soc. London, Ser. B*, **371**(1704), p. 20150390.

124. Liu, J., Dukes, I. and Hu, H. (2005). Novel mechatronics design for a robotic fish, in *2005 IEEE/RSJ International Conference on Intelligent Robots and Systems*, pp. 807–812.

125. Low, K. H. and Willy, A. (2006). Biomimetic motion planning of an undulating robotic fish fin, *J. Vib. Control*, **12**(12), pp. 1337–1359.

126. Low, K. H., Zhou, C. and Zhong, Y. (2009). Gait planning for steady swimming control of biomimetic fish robots, *Adv. Rob.*, **23**(7–8), pp. 805–829.

127. Lu, H., Lua, K. B., Lee, Y. J., Lim, T. T. and Yeo, K. S. (2016). Ground effect on the aerodynamics of three-dimensional hovering wings, *Bioinspir. Biomim.*, **11**(6), p. 066003.

128. Lua, K. B., Lai, K. C., Lim, T. T. and Yep, K. S. (2010). On the aerodynamic characteristics of hovering rigid and flexible hawkmoth-like wings, *Exp. Fluids*, **49**, pp. 1263–1291.

129. Ma, K., Chirarattananon, P., Fuller, S. and Wood, R. (2013). Controlled flight of a biologically inspired, insect-scale robot, *Science*, pp. 603–607.

130. Manfredi, L., Assaf, T., Mintchev, S., Marrazza, S., Capantini, L., Orofino, S., Ascari, L., Grillner, S., Wallén, P., Ekeberg, Ö., Stefanini, C. and Dario,

P. (2013). A bioinspired autonomous swimming robot as a tool for studying goal-directed locomotion, *Biol. Cybern.*, **107**(5), pp. 513–527.

131. Masoomi, S. F., Gutschmidt, S., Chen, X. and Sellier, M. (2015). The kinematics and dynamics of undulatory motion of a tunamimetic robot, *Int. J. Adv. Rob. Syst.*, **12**(7), p. 83.

132. Maxworthy, T. (2007). The formation and maintenance of a leading-edge vortex during the forward motion of an animal wing, *J. Fluid. Mech.*, **587**, pp. 471–475.

133. Mistick, E. A., Mountcastle, A. M. and Combes, S. A. (2016). Wing flexibility improves bumblebee flight stability, *J. Exp. Biol.*, **219**, pp. 3384–3390.

134. Mountcastle, A. M. and Combes, S. A. (2013). Wing flexibility enhances load-lifting capacity in bumblebees, *Proc. R. Soc. London, Ser. B*, **280**, p. 20130531.

135. Müller, U. K., van den Boogaart, J. G. M. and van Leeuwen, J. L. (2007). Flow patterns of larval fish: undulatory swimming in the intermediate flow regime, *J. Exp. Biol.*, **211**(2), pp. 196–205.

136. Nakata, T. and Liu, H. (2012). Aerodynamic performance of a hovering hawkmoth with flexible wings: a computational approach, *Proc. R. Soc. London, Ser. B*, **279**, pp. 722–731.

137. Nakata, T., Liu, H., Tanaka, Y., Nishihashi, N., Wang, X. and Sato, A. (2011). Aerodynamics of a bio-inspired flexible flapping-wing micro air vehicle, *Bioinspir. Biomim.*, **6**(4), p. 045002.

138. Noda, R., Nakata, T. and Liu, H. (2014). Effects of wing deformation on aerodynamic performance of a revolving insect wing, *Acta Mech. Sin.*, **30**(6), pp. 819–827.

139. Ortega-Jimenez, V. M., Badger, M., Wang, H. and Dudley, R. (2016). Into rude air: hummingbird flight performance in variable aerial environments, *Phil. Trans. R. Soc. B*, **371**(1704).

140. Ortega-Jimenez, V. M., Greeter, J. S. M., Mittal, R. and Hedrick, T. L. (2013). Hawkmoth flight stability in turbulent vortex streets, *J. Exp. Biol.*, **216**(24), pp. 4567–4579.

141. Ortega-Jimenez, V. M., Mittal, R. and Hedrick, T. L. (2014a). Hawkmoth flight performance in tornado-like whirlwind vortices, *Bioinspir. Biomim.*, **9**(2), p. 025003.

142. Ortega-Jimenez, V. M., Sapir, N., Wolf, M., Variano, E. A. and Dudley, R. (2014b). Into turbulent air: size-dependent effects of von Kármán vortex streets on hummingbird flight kinematics and energetics, *Proc. R. Soc. London, Ser. B*, **281**(1783), p. 20140180.

143. Ozen, C. A. and Rockwell, D. (2012). Flow structure on a rotating plate, *Exp. Fluids*, **52**(1), pp. 207–223.

144. Park, S.-J., Gazzola, M., Park, K. S., Park, S., Di Santo, V., Blevins, E. L., Lind, J. U., Campbell, P. H., Dauth, S., Capulli, A. K., Pasqualini, F. S., Ahn, S., Cho, A., Yuan, H., Maoz, B. M., Vijaykumar, R., Choi, J.-W., Deisseroth, K., Lauder, G. V., Mahadevan, L. and Parker, K. K. (2016). Phototactic guidance of a tissue-engineered soft-robotic ray, *Science*, **353**(6295), pp. 158–162.

145. Percin, M. and van Oudheusden, B. W. (2015). Three-dimensional flow structures and unsteady forces on pitching and surging revolving at plates, *Exp. Fluids*, **56**(2), pp. 1–19.

146. Phillips, N., Knowles, K. and Bomphrey, R. J. (2015). The effect of aspect ratio on the leading-edge vortex over an insect-like aping wing, *Bioinspir. Biomim.*, **10**(5), p. 056020.

147. Phillips, N., Knowles, K. and Bomphrey, R. J. (2017). Petiolate wings: effects on the leading-edge vortex in flapping flight, *Interface Focus*, **7**(1), p. 20160084.

148. Raj, A. and Thakur, A. (2016). Fish-inspired robots: design, sensing, actuation, and autonomy-a review of research, *Bioinspir. Biomim.*, **11**(3), p. 031001.

149. Ramananarivo, S., Godoy-Diana, R. and Thiria, B. (2011). Rather than resonance, flapping wing flyers may play on aerodynamics to improve performance, *Proc. Natl. Acad. Sci. U. S. A.*, **108**, pp. 5964–5969.

150. Ramananarivo, S., Godoy-Diana, R. and Thiria, B. (2013). Passive elastic mechanism to mimic fish-muscle action in anguilliform swimming, *J. R. Soc. Interface*, **10**(88).

151. Ravi, S., Crall, J. D., Fisher, A. M. and Combes, S. A. (2013). Rolling with the flow: bumblebees flying in unsteady wakes, *J. Exp. Biol.*, **216**(22), pp. 4299–4309.

152. Ravi, S., Crall, J. D., McNeilly, L., Gagliardi, S. F., Biewener, A. A. and Combes, S. A. (2015). Hummingbird flight stability and control in freestream turbulent winds, *J. Exp. Biol.*, **218**(9), pp. 1444–1452.

153. Ravi, S., Kolomenskiy, D., Engels, T., Schneider, K., Wang, C., Sesterhenn, J. and Liu, H. (2016). Bumblebees minimize control challenges by combining active and passive modes in unsteady winds, *Sci. Rep.*, **6**, p. 35043.

154. Reddy, G., Celani, A., Sejnowski, T. J. and Vergassola, M. (2016). Learning to soar in turbulent environments, *Proc. Natl. Acad. Sci. U. S. A.*, **113**(33), pp. E4877–E4884.

155. Rees, C. J. (1975). Form and function in corrugated insect wings, *Nature*, **256**, pp. 200–203.

156. Ristroph, L., Bergou, A. J., Ristroph, G., Coumes, K., Berman, G. J., Guckenheimer, J., Wang, Z. J. and Cohen, I. (2010). Discovering the flight autostabilizer of fruit flies by inducing aerial stumbles, *Proc. Natl. Acad. Sci. U. S. A.*, **107**(11), pp. 4820–4824.

157. Roche, D. G., Taylor, M. K., Binning, S. A., Johansen, J. L., Domenici, P. and Steffensen, J. F. (2014). Unsteady flow affects swimming energetics in a labriform fish (cymatogaster aggregata), *J. Exp. Biol.*, **217**(3), pp. 414–422.

158. Rossi, C., Coral, W., Colorado, J. and Barrientos, A. (2011). A motor-less and gear-less bio-mimetic robotic fish design, in *2011 IEEE International Conference on Robotics and Automation*, pp. 3646–3651.

159. Ryuh, Y.-S., Yang, G.-H., Liu, J. and Hu, H. (2015). A school of robotic fish for mariculture monitoring in the sea coast, *J. Bionic Eng.*, **12**(1), pp. 37–46.

160. Sfakiotakis, M., Lane, D. M. and Davies, J. B. C. (1999). Review of fish swimming modes for aquatic locomotion, *IEEE J. Oceanic Eng.*, **24**(2), pp. 237–252.

161. Shao, J., Wang, L. and Yu, J. (2007). Development of multiple robotic fish cooperation platform, *Int. J. Syst. Sci.*, **38**(3), pp. 257–268.

162. Shirgaonkar, A. A., Curet, O. M., Patankar, N. A. and MacIver, M. A. (2008). The hydrodynamics of ribbon-fin propulsion during impulsive motion, *J. Exp. Biol.*, **211**(21), pp. 3490–3503.

163. Shoele, K. and Zhu, Q. (2010). Numerical simulation of a pectoral fin during labriform swimming, *J. Exp. Biol.*, **213**(12), pp. 2038–2047.

164. Shyy, W., Aono, H., Chimakurthi, S., Trizila, P., Kang, C.-K., Cesnik, C. and Liu, H. (2010). Recent progress in flapping wing aerodynamics and aeroelasticity, *Prog. Aerosp. Sci.*, **46**, pp. 284–327.

165. Shyy, W., Kang, C.-K., Chirarattananon, P., Ravi, S. and Liu, H. (2016). Aerodynamics, sensing and control of insect-scale flapping wing flight, *Proc. R. Soc. London, Ser. A*, **472**, p. 20150712.

166. Shyy, W., Lian, Y., Tang, J., Viieru, D. and Liu, H. (2007). *Aerodynamics of Low Reynolds Number Flyers* (Cambridge University Press).

167. Sims, T. W., Palazotto, A. N. and Norris, A. (2010). A structural dynamic analysis of a manduca sexta forewing, *Int. J. Micro Air Veh.*, **2**, pp. 119–140.

168. Sitorus, P. E., Nazaruddin, Y. Y., Leksono, E. and Budiyono, A. (2009). Design and implementation of paired pectoral fins locomotion of labriform fish applied to a fish robot, *J. Bionic Eng.*, **6**(1), pp. 37–45.

169. Sridhar, M. and Kang, C.-K. (2015). Aerodynamic performance of two-dimensional, chordwise flexible flapping wings at fruit fly scale in hover flight. *Bioinspir. Biomim.*, **10**, p. 036007.

170. Standen, E. M. and Lauder, G. V. (2007). Hydrodynamic function of dorsal and anal fins in brook trout (salvelinus fontinalis), *J. Exp. Biol.*, **210**(2), pp. 325–339.

171. Steppan, S. J. (2000). Flexural stiffness patterns of buttery wings (Papilionoidea), *J. Res. Lepid.*, **35**, pp. 61–77.

172. Sunada, S., Zeng, L. and Kawachi, K. (1998). The relationship between dragonfly wing structure and torsional deformation, *J. Theor. Biol.*, **193**, pp. 39–45.

173. Taguchi, M. and Liao, J. C. (2011). Rainbow trout consume less oxygen in turbulence: the energetics of swimming behaviors at different speeds, *J. Exp. Biol.*, **214**(9), pp. 1428–1436.

174. Tangorra, J. L., Davidson, S. N., Hunter, I. W., Madden, P. G. A., Lauder, G. V., Dong, H., Bozkurttas, M. and Mittal, R. (2007). The development of a biologically inspired propulsor for unmanned underwater vehicles, *IEEE J. Oceanic Eng.*, **32**(3), pp. 533–550.

175. Taylor, G. (1951). Analysis of the swimming of microscopic organisms, *Proc. R. Soc. London, Ser. A*, **209**(1099), pp. 447–461.

176. Toming, G., Chambers, L. D. and Kruusmaa, M. (2014). Experimental study of hydrodynamic forces acting on artificial fish in a von kármán vortex street, *Underwater Technol.*, **32**(2), pp. 81–91.

177. Tritico, H. M. and Cotel, A. J. (2010). The effects of turbulent eddies on the stability and critical swimming speed of creek chub (semotilus atromaculatus), *J. Exp. Biol.*, **213**(13), pp. 2284–2293.

178. Trivedi, D., Rahn, C. D., Kier, W. M. and Walker, I. D. (2008). Soft robotics: biological inspiration, state of the art, and future research, *Appl. Bionics Biomech.*, **5**(3), pp. 99–117.

179. Tytell, E. D. (2006). Median fin function in bluegill sunfish lepomis macrochirus: streamwise vortex structure during steady swimming, *J. Exp. Biol.*, **209**(8), pp. 1516–1534.

180. Tytell, E. D. and Lauder, G. V. (2008). Hydrodynamics of the escape response in bluegill sunfish, lepomis macrochirus, *J. Exp. Biol.*, **211**(21), pp. 3359–3369.

181. Usherwood, J. W. and Ellington, C. P. (2002). The aerodynamics of revolving wings. I. Model hawkmoth wing, *J. Exp. Biol.*, **205**, pp. 1547–1564.

182. Van Wassenbergh, S., van Manen, K., Marcroft, T. A., Alfaro, M. E. and Stamhuis, E. J. (2014). Boxfish swimming paradox resolved: forces by the flow of water around the body promote manoeuvrability, *J. R. Soc. Interface*, **12**(103).

183. Vance, J. T., Faruque, I. and Humbert, J. S. (2013). Kinematic strategies for mitigating gust perturbations in insects, *Bioinspir. Biomim.*, **8**(1), p. 016004.

184. Venturelli, R., Akanyeti, O., Visentin, F., Ježov, J., Chambers, L. D., Toming, G., Brown, J., Kruusmaa, M., Megill, W. M. and Fiorini, P. (2012). Hydrodynamic pressure sensing with an artificial lateral line in steady and unsteady flows, *Bioinspir. Biomim.*, **7**(3), p. 036004.

185. Voesenek, C. J., Pieters, R. P. M. and van Leeuwen, J. L. (2016). Automated reconstruction of three-dimensional fish motion, forces, and torques, *PLoS One*, **11**(1), pp. 1–17.

186. Walker, J. A. (2000). Does a rigid body limit maneuverability? *J. Exp. Biol.*, **203**(220, pp. 3391–3396.

187. Walker, S. M., Thomas, A. L. R. and Taylor, G. K. (2009). Photogrammetric reconstruction of high-resolution surface topographies and deformable wing kinematics of tethered locusts and free-flying hoverflies, *J. R. Soc. Interface*, **6**, pp. 351–366.

188. Wang, W. and Xie, G. (2014). Cpg-based locomotion controller design for a boxfish-like robot, *Int. J. Adv. Rob. Syst.*, **11**(6), p. 87.

189. Wang, Y., Tan, J. and Zhao, D. (2015). Design and experiment on a biomimetic robotic fish inspired by freshwater stingray, *J. Bionic Eng.*, **12**(2), pp. 204–216.

190. Wang, Z., Wang, Y., Li, J. and Hang, G. (2009). A micro biomimetic manta ray robot fish actuated by sma, in *2009 IEEE International Conference on Robotics and Biomimetics (ROBIO)*, pp. 1809–1813.

191. Webb, P. W. (1984). Form and function in fish swimming, *Sci. Am.*, **251**, pp. 58–68.

192. Weihs, D. (2011). Hydromechanics of fish schooling, *Nature*, **241**(5387), pp. 290–291.

193. Weis-Fogh, T. (1960). A rubber-like protein in insect cuticle, *J. Exp. Biol.*, **37**, pp. 889–907.

194. Whitney, J. and Wood, R. (2012). Conceptual design of flapping-wing micro air vehicles, *Bioinspir. Biomim.*, **7**, p. 036001.

195. Whitney, J. P. and Wood, R. J. (2010). Aeromechanics of passive rotation in flapping flight, *J. Fluid Mech.*, **660**, pp. 197–220.

196. Wojcik, C. J. and Buchholz, J. H. J. (2014). Vorticity transport in the leading-edge vortex on a rotating blade, *J. Fluid. Mech.*, **743**, pp. 249–261.

197. Wolfgang, M. J., Anderson, J. M., Grosenbaugh, M. A., Yue, D. K. and Triantafyllou, M. S. (1999). Near-body flow dynamics in swimming fish, *J. Exp. Biol.*, **202**(17), pp. 2303–2327.

198. Wootton, R. J. (1981). Support and deformability in insect wings, *J. Zool.*, **193**, pp. 447–468.

199. Wootton, R. J. (1992). Functional morphology of insect wings, *Annu. Rev. Entomol.*, **37**, pp. 113–140.

200. Wu, T. (2011). Fish swimming and bird/insect flight, *Annu. Rev. Fluid Mech.*, **43**, pp. 25–58.

201. Yanase, K., Herbert, N. A. and Montgomery, J. C. (2012). Disrupted flow sensing impairs hydrodynamic performance and increases the metabolic cost of swimming in the yellowtail kingfish, seriola lalandi, *J. Exp. Biol.*, **215**(22), pp. 3944–3954.

202. Yin, B. and Luo, H. (2010). Effect of wing inertia on hovering performance of flexible flapping wings, *Phys. Fluids*, **22**, p. 111902.

203. Young, J., Walker, S. M., Bomphrey, R. J., Taylor, G. K. and Thomas, A. L. R. (2009). Details of insect wing design and deformation enhance aerodynamic function and flight efficiency, *Science*, **325**, pp. 1549–1552.

204. Yu, C.-L., Ting, S.-C., Yeh, M.-K. and Yang, J.-T. (2011). Three-dimensional numerical simulation of hydrodynamic interactions between pectoral-fin vortices and body undulation in a swimming fish, *Phys. Fluids*, **23**(9), p. 091901.

205. Yu, S., Ma, S., Li, B. and Wang, Y. (2009). An amphibious snakelike robot: design and motion experiments on ground and in water, in *2009 International Conference on Information and Automation*, pp. 500–505.

206. Zhao, L., Huang, Q., Deng, X. and Sane, S. P. (2010). Aerodynamic effects of flexibility in flapping wings, *J. R. Soc. Interface*, **7**, pp. 485–497.

207. Zheng, L., Hedrick, T. L. and Mittal, R. (2012). Time-varying wing-twist improves aerodynamic efficiency of forward flight in butteries, *PLoS One*, **8**(1), p. e53060.

Chapter 4

Shape-Tunable Wrinkles Can Switch Frictional Properties

Takuya Ohzono and Kosuke Suzuki
Research Institute for Sustainable Chemistry, National Institute of Advanced Industrial Science and Technology (AIST), 1-1-1 Higashi, Tsukuba, Ibaraki 305-8565, Japan
ohzono-takuya@aist.go.jp

4.1 Introduction

4.1.1 Friction at Soft Structured Interfaces

Living matters such as animals, insects, and plants exhibit a wide variety of soft textured surfaces. The asperities of the textured surfaces contribute to their tribological properties, for example, adhesion [1–6], lubrication [7], and friction (on a finger pad) [8–10]. In particular, the asperities of the soft textured surface deform greatly under sliding friction, and the deformation can be continuous, periodic, or intermittent. The dynamics of the deformation should affect the fluctuations in the friction force and its average. An

Industrial Biomimetics
Edited by Akihiro Miyauchi and Masatsugu Shimomura

ISBN 978-981-4800-07-5 (Hardcover), 978-0-429-05883-7 (eBook)
www.jennystanford.com

understanding of the relationship between the deformation of the structured surface and friction provides useful information for engineering friction at soft interfaces. However, it is difficult to understand directly the mechanism of friction on soft living surfaces because it is associated with the time-dependent deformation of complicated hierarchical structures. Therefore, as a first step, it would be important to study the friction using a simplified surface structure, which makes the analysis easier.

In addition to the actively studied friction between hard asperities and soft flat surfaces [11–17], the effects of various artificial soft-and-deformable asperities on frictional phenomena, for example, contact, deformation, adhesion, and friction, have also been investigated [18–35]. Soft textured rubbers can deform locally, and the total area and number of the contacts depend nonlinearly on the applied normal load [20, 21, 24–26], which is the main difference from a surface without asperity. To understand contact problems correctly, some theoretical approaches [18, 19, 22, 23, 35] based on, for example, the Johnson–Kendall–Roberts framework [19], are necessary. Regarding friction on a soft textured surface, Rand and Crosby reported interesting results when they used aligned wrinkles [36–39] imprinted on an elastomer, polydimethylsiloxane (PDMS), surface [29]. They reported that the averaged friction force was reduced and that the oscillation of the friction force generated on a smooth PDMS surface by the formation of Schallamach waves [40, 41] was suppressed on the wrinkled surface. In their experiments, the radius of the hemispherical indenter was roughly 100 times larger than the wavelength of the wrinkles, λ. However, the frictional dynamics, in general, should depend on the relative length between the radius of the indenter and the typical length of the topographic structure of the counter surface. Thus, such a series of studies are required to overview the effects of the relative length at the frictional interface and to illuminate how the spatiotemporal deformation of soft textures contributes to the friction.

To conduct such an experiment, we can use shape-tunable wrinkles [42]; their sinusoidal wavy surface structure with a specific periodicity is suitable to investigate how the topography affects the frictional dynamics and depth and direction of the wrinkle grooves can be reversibly controlled by applying a strain [24, 36–39, 43–46]. These properties have been applied to shape liquids at small scales by controlling capillary phenomena through changes in the microgroove shape [24, 39, 46–48], to control the liquid crystal alignment [47, 48], and to control the optical properties [42]. With the shape-tunable wrinkles, the shape-dependent frictional forces are expected to be controlled simply by the applied strain, and the tunability of the friction may be applicable to, for example, touch-screen devices [49] and tunable grips. In the next section, the basic properties of the shape-tunable wrinkles are briefly described, and then, a series of frictional studies on the a few different wrinkles with their fabrication methods follows.

4.1.2 Shape-Tunable Wrinkles

In this section, the basic properties of wrinkles are reviewed. We here use the term "wrinkles" as a surface with periodic undulations formed through mechanical instability, such as Euler's buckling. Euler's buckling is a phenomenon that occurs during the bending of a plate or a beam under lateral compressive strain. The simplest form of a free plate after buckling under lateral uniaxial stress is approximately half a sine wave with a spatial wavelength double that of the total length of the plate in the stress direction. When the plate is supported by a compliant elastic or viscoelastic substrate, the resultant spatial wavelength (periodicity) of the wrinkles is reduced and depends on the relative modulus of the plate and soft substrate (Fig. 4. 1). This is expressed as $\lambda \propto h\,(E_f/E_s)^{1/3}$, where λ is the wavelength, h is the thickness of the plate, and E_f and E_s are the Young's modulus of the plate and the soft substrate (which is thicker than the plate), respectively [36–39].

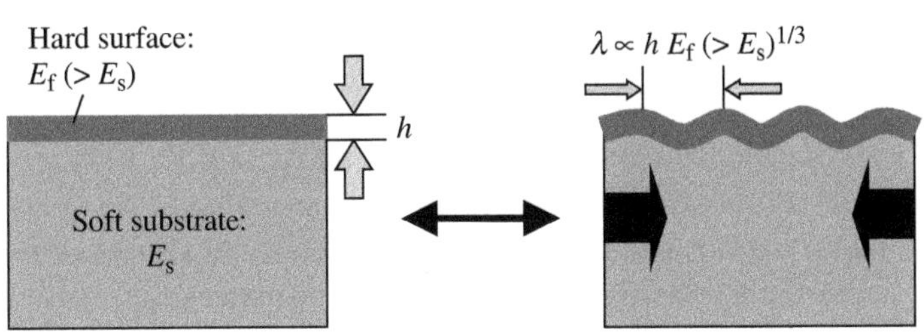

Figure 4.1 Schematics for the wrinkle system. The structure required for wrinkle formation. The hard surface layer is supported by a soft elastic substrate. Wrinkles form on this structure under lateral compressive strain.

The wrinkles with various length scales can be found in nature. The wrinkles on the surface of an animal's body are a familiar example; the relatively hard skin supported by the soft inner tissue can wrinkle under compressive stress. The surface of a paved road may wrinkle because of the expansion of the top hard surface as it heats up in strong sunlight. A similar phenomenon occurs in oil paintings; environmental damage causes the hard surface to expand or the base to shrink. Wrinkle formation is sometimes an irreversible process, such as plastic deformation of a hard skin layer and delamination of a hard layer from a soft substrate (Fig. 4.1b). Although irreversible processes are important in practice, in this study we will consider a simple system, where they are considered to be minimal.

Microwrinkle systems have been extensively studied in recent years [36-39, 42–48]. Wrinkles with macroscopic wavelengths may not be technologically useful; however, microwrinkles with wavelengths in the nanometer-to-micrometer order have potential utility in various fields of technology. These surface microstructures are spontaneously formed through a mechanical self-organization process without conventional lithographic techniques and have many potential practical applications in optical devices, microfluidic devices, cell culture substrates, and thin-film devices such as flexible electronics.

Microwrinkles are fabricated using a thin film, typically several tens of nanometers thick, which is supported by a relatively soft substrate. The wavelength of the wrinkles is determined by the thickness of the hard film and the mechanical properties of the hard film and the soft substrate. Therefore, in principle microwrinkle structures can be produced in various kinds of technologically important materials (Fig. 4.2). There are four main methods for film fabrication:

- Direct deposition of metals, such as Au and Pt, onto a soft elastic substrate is the most common method. Dip coating, spin coating, and various plating methods can also be used for the deposition of inorganic and organic materials.
- A hard skin layer that acts as the thin film can be produced by surface modification of silicone rubber by mild oxygen plasma treatment. Active oxygen species generated by the

plasma treatment transform the soft silicone surface into a hard, silica-like skin through chemical degradation.

- The layer-by-layer method can be used to form a thin film of ionic polymers [50].
- Self-supporting thin organic and inorganic films, which are prepared separately [51], can be transferred onto a soft substrate to form heterostructures.

All these methods allow the film thickness to be controlled on the nanometer scale, and thus the wavelength of the microwrinkles is easily controlled; the typical range is between a few hundred nanometers and hundred micrometers. The wrinkles are actually induced by the exertion of lateral compressive strain or stress.

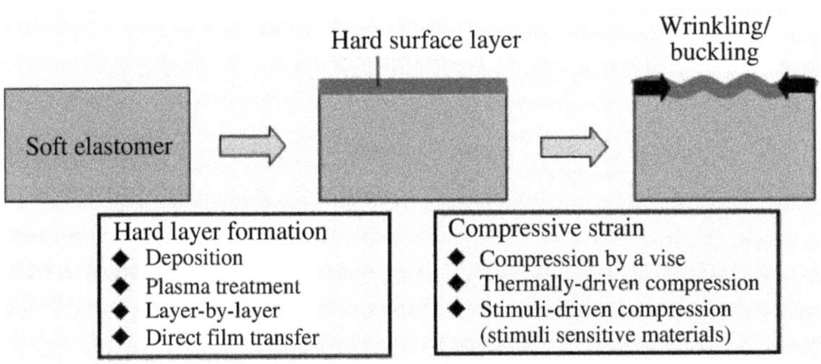

Figure 4.2 Several methods for preparing microwrinkles.

As mentioned before, the wavelength is determined by the mechanical properties of the hard film and the soft substrate and by the thickness of the hard film. On the other hand, the pattern of the wrinkles depends on the 2D compressive strain or stress state. Uniaxial strain creates simple patterns of stripes composed of straight crest/groove lines, whereas isotropic strain produces complex labyrinthine or herringbone patterns (Fig. 4.3). Therefore, in principle it is possible to change the orientation of the wrinkles by altering the externally applied strain; the small topographic structures may be reversibly changed by the addition of a certain type of strain. If we can reversibly control the surface microstructure, this dynamic system could be applied to optical devices, microfluidic devices, cell culture devices, and thin-film devices. In this chapter, the tunability of friction exploiting the shape tunability of a wrinkled surface is addressed.

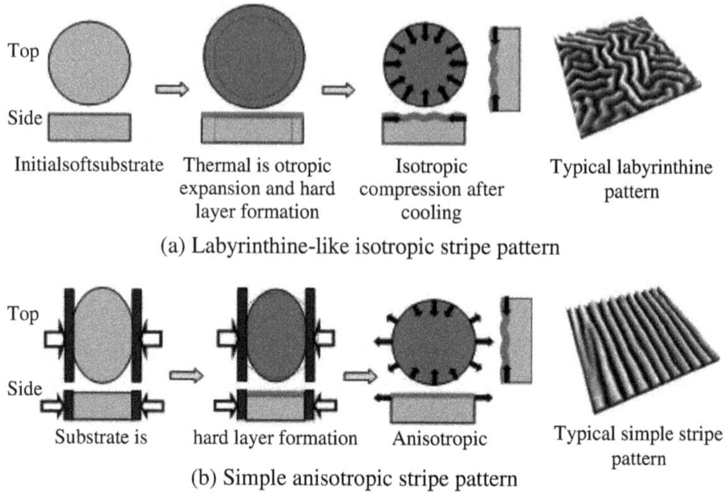

Figure 4.3 Strain-anisotropy-dependent stripe patterns of wrinkles.

In summary, we have briefly reviewed shape-tunable microwrinkles. By changing the anisotropy and absolute value of the applied strain, various transformations of the microwrinkle stripe patterns are possible. Other characteristic features of the transformation are that the wavelength remains almost constant and that the process is reversible with some hysteresis. The amplitude, A, of the microwrinkles, which is proportional to the height difference between the crest and the groove bottom, also can be controlled by the strain, s, according to the relation $A \propto s^{1/2}$ beyond the critical buckling strain. In the next section, the application of various types of shape-tunable microwrinkles for manipulating friction will be described.

4.2 Friction on Wrinkled Surfaces

Working with the frictional properties [52–54] exhibited by soft materials [55, 56], such as rubber and plastics, has demonstrated great practical importance in many applications, including seals, windshield wipers, and grips on tools, robot hands, and sports equipment. Many surface structures of rubbery materials have been proposed [57–59] to enhance both static friction and adhesion. The forms of natural contacting surfaces provide the inspiration for some of these structures. Some possess fibrillar structures and their analogs that enlarge the real contact area and yield enhanced static friction. Some practical applications of these surfaces depend on simultaneously considering the efficient detachment process. Moreover, practical surfaces, such as grips, show a fundamental need for on-demand tuning of the static, dynamic frictions and adhesion forces. Such tunability may be realized by systems in which some external triggers at a constant normal load vary the friction force (Fig. 4.5). To design such systems, light-responsive polymers, which modify the elasticity of the interface [60], or strain-responsive structuring of the surface, that is, shape-tunable wrinkles [42-47], may be effective to change frictional properties. In the present chapter, we focus on the latter system and briefly review the recent progress on them.

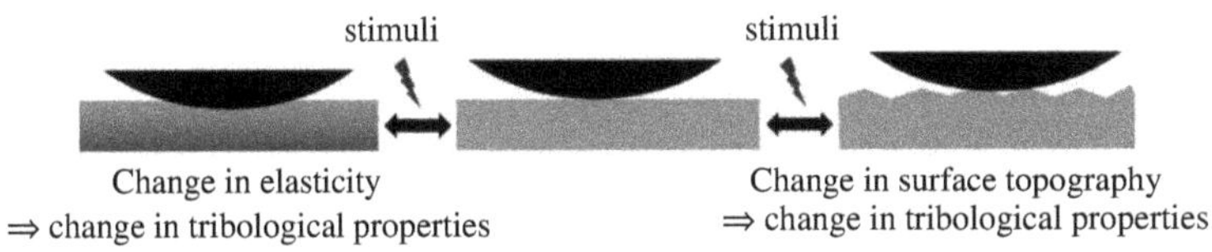

Figure 4.4 Strain-anisotropy-dependent stripe patterns of wrinkles.

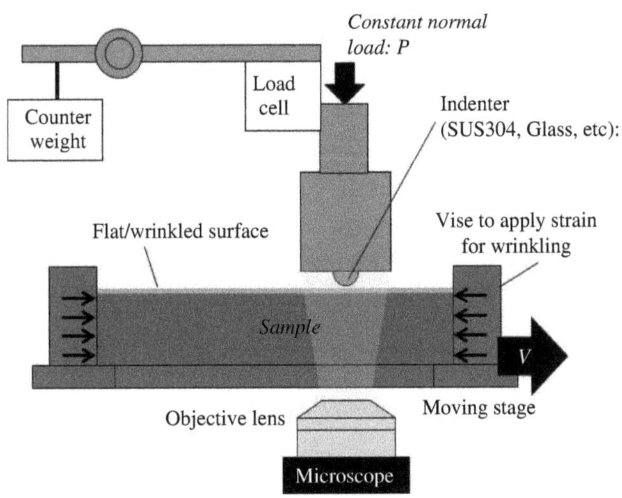

Figure 4.5 Schematic illustration of our experimental setup for friction measurement and in situ observation. Adapted with permission from Ref. [61]. Copyright (2014) American Chemical Society. https://pubs.acs.org/doi/abs/10.1021%2Fam5010738 . Further permissions related to the material excerpted should be directed to the ACS.

4.2.1 Friction Force Measurements

The friction force was mainly measured by a pin-on-plate friction tester equipped with an inverted reflection microscope. The typical experimental setup is shown in Fig. 4.5. For the friction force measurements, we used a glass or stainless-steel indenter with a hemispherical shape with different radius, R, which was connected to a beam-type load cell. The sample held by a vise was mounted on a linear sliding stage for frictional sliding. The alignment of the wrinkles was set either parallel or perpendicular to the sliding direction. Microscopic images of the frictional interface and the surrounding area were continuously recorded using a video camera. In some experiments, their time series of the recoded images were reconstructed into a spatiotemporal image to investigate the temporal evolution of the deformation of the wrinkled surface.

4.2.2 Preparation of Polyimide Wrinkles

Shape-tunable wrinkles were prepared by attaching a thin polyimide film (Kapton®, thickness of 12.5 μm, elastic modulus of 2.5 GPa) onto an elastomeric PDMS substrate. To induce wrinkles, a uniaxial compressive strain (s: 0%–7.5%) was applied to a sample using a home-made vise.

When the applied strain exceeded 1.4%, the sample surface buckled into wrinkles. The spatial wavelength of the wrinkles, λ, decreased slightly as the strain increased further. In addition, the amplitude h exhibited an inverse-quadratic increase from 43 μm to 103 μm (Fig. 4.6 a). Because the strain dependence of λ was small, the aspect ratio (h/λ) also showed inverse-quadratic increase with respect to s and changed from 0.073 to 0.185 (Fig. 4.6b). Figure 4.6c,d shows reflection microscopic images of the sample surfaces. The flat surface (at s = 0%) showed no pattern (Fig. 4.6c). In contrast, the wrinkled surface (at s = 7.5%) showed periodic bright lines, each of which corresponds alternately to either the bottom or top of the wrinkled surface (Fig. 4.6d). The shape tunability shown in Fig. 4.6 is possible on the polyimide wrinkles, which will be used as the shape-tunable surface to change the friction in the next two subsections.

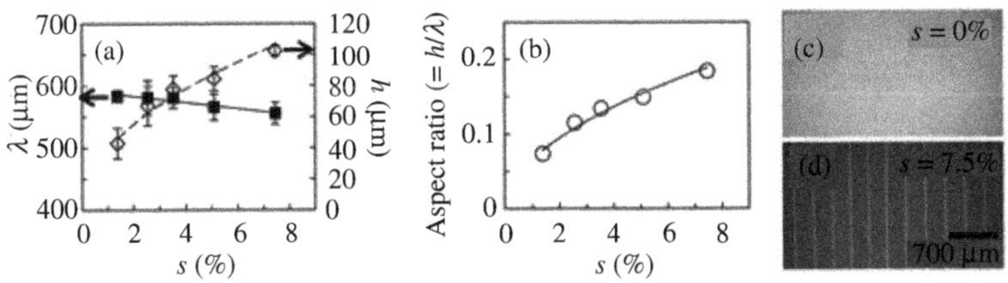

Figure 4.6 Characteristic lengths of shape-tunable wrinkles and optical images. (a) Plots of wavelength (◊) and amplitude h (■) with respect to applied strain s. (b) Plot of aspect ratio h/λ with respect to s. Microscopic images of (c) flat and (d) wrinkled surfaces. Adapted with permission from Ref. [61]. Copyright (2014) American Chemical Society. https://pubs.acs.org/doi/abs/10.1021%2Fam5010738 . Further permissions related to the material excerpted should be directed to the ACS.

4.2.3 Friction with a Small Indenter on Polyimide Wrinkles

Here, the friction on polyimide wrinkles with the wavelength $\lambda \approx 0.56$ mm using an indenter ball with a small radius (R = 1.0 mm) [61] is briefly reviewed; $R \approx \lambda$. Figure 4.7 shows typical friction profiles and microscopic images observed during sliding at a constant sliding speed of 0.3 mm/s at various normal loads and strains on wrinkles that were aligned perpendicular to the sliding direction. The friction force oscillated on the wrinkled surfaces with a periodicity identical to that of the wrinkles independent of the normal loads and aspect ratio of wrinkles. In addition, the friction force was higher at a higher normal load, and the amplitude of the oscillation was larger at a higher normal load and/or larger strain (Fig. 4.7a). The periodic oscillation of the friction force was not observed on the flat surfaces (s = 0%); this was also the result when the wrinkles were aligned parallel to the sliding direction. This also suggests that no observable Schallamach wave developed in our system. In situ microscopic images taken during sliding between the surface and the indenter show that the deformation of the wrinkles in contact with the indenter depends on the applied normal loads and the strain (Fig. 4.7b). A larger deformation area was observed under a higher normal load. Under lower loads, one or two crests were deformed (Fig. 4.7b–v,vi,viii,ix)), and two or more crests were deformed under higher loads (Fig. 4.7b–ii, iii). As discussed later, the normal-load-dependent different modes of contact affect the oscillation of the friction. In the followings, the averaged friction forces are described.

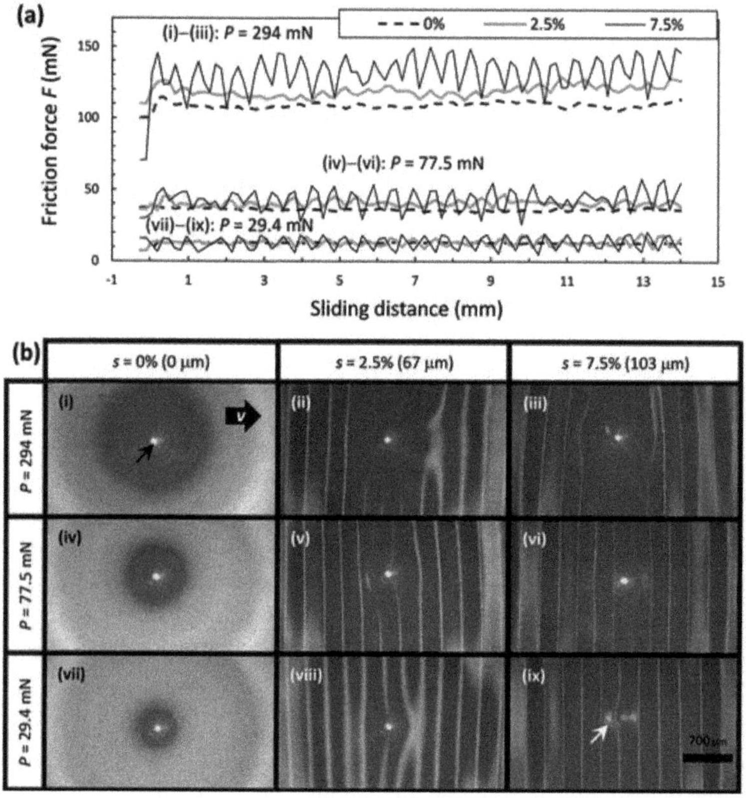

Figure 4.7 Typical friction profiles and snapshots of wrinkled and flat surfaces under friction. (a) Friction profiles of wrinkles aligned perpendicular to the sliding direction (gray bold lines: s = 2.5%, h = 67 μm; solid lines: s = 7.5%, h = 103 μm) and flat surfaces (broken lines). (b) In situ observation of friction surface under optical microscopic images; the applied normal loads are P = 294 mN for (i–iii), 77.5 mN for (iv–vi), and 29.4 mN for (vii–ix). The bold black arrow indicates direction of sliding of the sample surface with speed v = 0.3 mm/s. The black arrow in (b–i) indicates the center of the indenter, which remains in the images. The white arrow in (b–ix) indicates the bright area originating from enhanced multiple reflections when the center of the indenter is not in contact with the wrinkled surface. Adapted with permission from Ref. [61]. Copyright (2014) American Chemical Society. https://pubs.acs.org/doi/abs/10.1021%2Fam5010738 . Further permissions related to the material excerpted should be directed to the ACS.

The averaged friction force F_{av} depends almost linearly on the normal load (Fig. 4.8a). This linearity should be the result of a complicated interplay between the effects of the contact states and the deformation of wrinkles. Moreover, also the effect of the polyimide (PI) film might be involved in this scenario because the plots for the flat PI-capped PDMS also shows almost linear relationship instead of the power law one [50, 51], which is widely observed for the friction on the soft substrate (without the capping film). Nevertheless, from the results we can discuss the trend of the friction that depends on the surface states, (flat or wrinkled) and the sliding direction as follows.

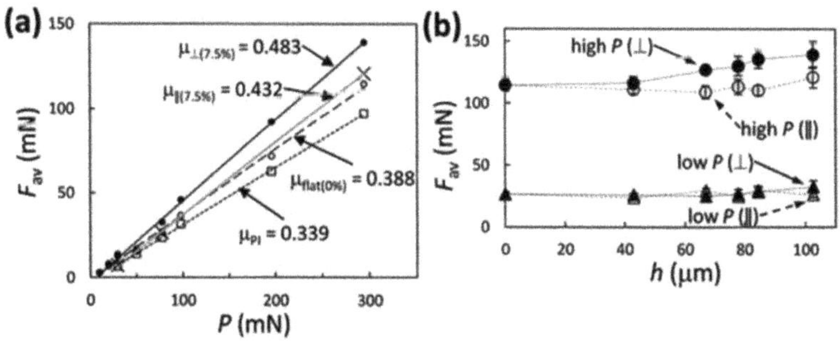

Figure 4.8 Averaged friction forces (F_{av}) on wrinkled surfaces. (a) F_{av} vs. P. Symbols: ○: flat surface (s = 0%); •: wrinkles (s = 7.5%, h = 103 μm) aligned perpendicular to the sliding direction (⊥); ×: those aligned parallel to the sliding direction (||); □: PI film on a glass plate. (b) F_{av} vs. h (amplitude of wrinkles) obtained under different sliding directions (closed symbols: ⊥; open ones: ||) and different normal loads (triangles: P = 77.5 mN; circles: P = 294 mN). Adapted with permission from Ref. [61]. Copyright (2014) American Chemical Society. https://pubs.acs.org/doi/abs/10.1021%2Fam5010738. Further permissions related to the material excerpted should be directed to the ACS.

The coefficient of friction (COF) on a PI film attached to a PDMS surface was larger than that on an identical PI film on a glass plate. The difference was likely due to the difference in the Young's moduli of the underlayers, which affect the frictional dissipation. A similar effect originating from the deformation of the underlayer was reported for rubber samples with different thicknesses [62]. When the surface was wrinkled, especially when the wrinkles were aligned perpendicular to the sliding direction, the COFs were slightly larger than that on the flat surface. This result seemingly contradicts the apparent reduction in the friction on wrinkles reported by Rand and Crosby [29]. However, considering that in our system, no Schallamach wave was observed even on the flat surface, it is reasonable to lose the reduction effect. Thus, it is also generally suggested that the effect of soft asperities on friction depends on whether Schallamach waves develop on a flat surface of the same material.

Figure 4.8b shows the effect of the applied strain, which controlled the amplitude of the wrinkles, on the averaged friction force. The data on the wrinkled surfaces aligned perpendicular or parallel to the sliding direction under two different applied normal loads is shown. When the applied normal load was small (77.5 mN), the averaged friction forces were almost constant, and the difference between the two alignment directions was small. When the applied normal load was high (~294 mN), the difference became larger. That is, the averaged friction force on the perpendicularly aligned wrinkles clearly increased with the strain, but that on the parallely aligned surface changed little. Note that various contact situations can be considered when the indenter slides in the parallel with the wrinkle grooves; the indenter is, for example, in contact with a top of the crest, with a bottom of the groove, and in between them. The data shown in Fig. 4.8b as open symbols is averaged values over the various contact situations.

Next, we qualitatively discuss the origin of the observed difference in the average friction forces shown in Fig. 4.8. The present frictional system involves three generic mechanisms: (i) the adhesion models at adjacent interfaces, (ii) the plowing models (or hysteretic friction model), and (iii) the Coulombic interlocking mechanism [63]. In the adhesion model, the energy dissipation process resulting from the shear of adhesive junctions in regions immediately adjacent to the interface is considered, which is a simple system of friction with small deformation on a flat surface. In the plowing model, the large deformation and plowing of the soft substrate and the viscoelastic energy dissipation are considered. In the Coulombic interlocking mechanism, the geometric effect of the slope of the nonflat and relatively rigid surface is considered. Since the averaged friction force is enhanced on the wrinkled surface, the Coulombic interlocking mechanism is most likely applied. However, the wrinkles are largely deformed/plowed even under the lower normal load. Thus, it makes difficult to conclude that the origin of the enhancement is purely the Coulombic interlocking mechanism. The enhancement is rather resulted from the nonadditive interplay between the three mechanisms, which remains to be studied with more detailed theoretical model.

Nevertheless, the effect of the wrinkled topography on the total friction force can be discussed in terms of the three mechanisms and roughly quantified using the experimental data shown in Fig. 4.8a as follows: (A) The friction on the PI film attached to a rigid glass slide corresponds to the contribution of (i) the adhesion model. (B) The friction on the PI film attached to the soft substrate (PDMS) includes (i) the adhesion model and (ii) the plowing model. Thus, the friction enhanced from the case (A) to case (B), which is the actual effect of the soft substrate on the averaged friction force, can be attributed to the contribution of the deformation/plowing of the substrate. (C) Finally, the friction on the wrinkled PI film includes (i–iii) the Coulombic interlocking mechanism. Thus, the Coulombic interlocking mechanism mixed with the deformation/plowing of the asperity of wrinkle topography can account for the friction enhanced from the case (B) to case (C), which is the actual effect of the wrinkles on the averaged friction force. On the average over the present range of the normal load, this contribution accounts for ~20% of total friction estimated from COFs shown in Fig. 4.8a; we can tune ~20% of the total friction purely by changing the wrinkle topography. Similarly, the contribution of the adhesion model at adjacent interface accounts for ~60% and the sum of those of other two mechanisms, the Coulombic interlocking and plowing accounts for ~40% of total friction on the wrinkled surface.

The present results indicate that the friction force can be controlled by changing the shape and/or alignment direction of the wrinkles. More importantly, the present shape-tunable wrinkles enable us to change both of them reversibly and repeatedly on one sample by changing the degree and/or anisotropy of the applied strain [42]. The mechanism of the present geometry-dependent friction, which includes the different frictional states caused by the applied strain and the alignment direction, must be correlated with the contact state between the deformable wrinkled surface and the indenter. For further detailed investigation, please refer to Ref. [61]. Although the friction increases on the wrinkled state with the present small ball as the indenter, the opposite phenomenon found with the low-curvature indenter will be shown in the next sections.

4.2.4 Friction with a Large Indenter on Polyimide Wrinkles

Here, the friction on polyimide wrinkles with wavelength $\lambda \approx 0.6$ mm using an indenter ball with a large radius ($R = 10$ mm) is briefly reviewed; $R >> \lambda$. As the results of this condition, the friction profiles show no oscillation corresponding to each wrinkle crests, which contrasts with the previous case shown in Section 4.2.3. They only show fluctuation originated from the inhomogeneity of the periodic wrinkles, such as depth distribution and topological defects in their stripe patterns [64]. Figure 4.9 shows that the averaged friction force F_{av} depends almost linearly on the normal load. In contrast to the previous case shown in Section 4.2.3, the COF on a flat surface ($s = 0\%$) was larger than that on the wrinkles ($s = 16\%$). This result suggests that the relative length between the periodicity of wrinkles and the radius of the indenter can alter the effect of the wrinkles on the friction. That is, the friction can be increased or decreased depending on the relative length.

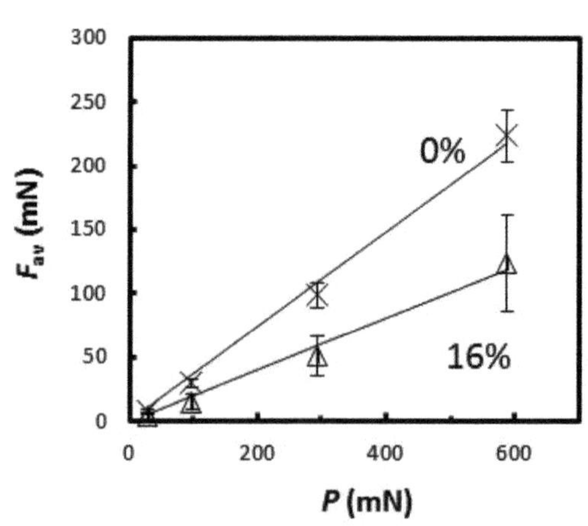

Figure 4.9 Averaged friction force F_{av} vs. applied normal load P. The error bars indicate standard deviations.

To estimate the reason for the decreased friction on the wrinkles with the present larger indenter, the contact states are observed, which is shown in Fig. 4.10. On the flat surface (s = 0%), the circular area appears as the main contact region. In contrast, on wrinkles (s = 16%) only the crest parts support the indenter. This main difference suggests that the friction is reduced because of the reduction of the contact area on the wrinkles. If we assume that the adhesion model for friction is applied to the present system, it is qualitatively reasonable results.

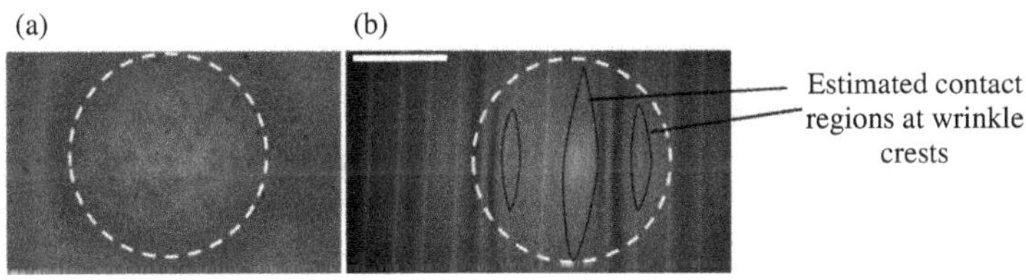

Figure 4.10 Frictional interface on transparent wrinkles under a constant load (98.0 mN). The indenter is sliding on (a) flat areas and (b) wrinkles. The white bar is 1 mm. White broken circles indicate the apparent contact region.

4.2.5 Preparation of Porous-Film-Embedded Wrinkles

In the following sections, new compositional designs for formation of wrinkles for tribological uses will be introduced and the tribological properties will be briefly reviewed. The structural designs proposed for wrinkling to improve mechanical durability by exploiting a porous polymer film or textile sheet embedded at the surface of an elastomer, which acts as a hard layer, buckles into wrinkles and effectively suppresses fatal failures such as delamination and cracking, are described with the frictional properties. In this section, the preparation of the porous-film-embedded wrinkles is firstly shown.

Generally, the top films (with Young's modulus typically over hundreds of megapascals) supported by a substrate cannot elastically accommodate a high strain, s (e.g., 0.1–0.2), resulting in crack formation or plastic deformation (yielding). Moreover, the adhesion between films and the substrate often cannot withstand the internal stress associated with the formation of wrinkles, leading to partial delamination known as buckling delamination. Such wrinkles are easily perturbed or destroyed by the application of large strains, repeated straining cycles, or mechanical rubbing.

Consequently, it is difficult to directly utilize such wrinkles for applications that require a certain level of mechanical durability, for example, tribological interfaces or switching elements with repeated structural changes. Thus, the use of wrinkles has been limited to an easy-to-make template for micropatterns. Thus, for use in tribological tests, patterns imprinted into other mechanically robust materials are employed; however, these materials do not possess the unique shape tunability of wrinkles, and this limits the controllability of tribological properties.

Therefore, a new design for a layered wrinkling structure with improved mechanical durability achieved by exploiting a porous polymer film embedded at an elastomer surface has been proposed. The porous film, which possesses 2D hexagonal symmetry and pillars (or spikes) [65] protruding in an out-of-plane direction, acts as a hard layer that buckles into wrinkles.

The porous polymer film to be embedded in a PDMS substrate is processed from a self-organized honeycomb-like porous film [65] and prepared on a glass substrate [66]. A pre-cured PDMS fluid is poured on the porous film and the film-embedded PDMS is peeled off from the glass substrate after curing. Although some pillar apexes are exposed to air at the surface (Fig. 4.11a), most parts of the porous film are buried within PDMS. Thus, the top surface of the sample is almost flat except the pillar locations, which show a slight depression of 50–200 nm (Fig. 4.11c,e). When a uniaxial strain (s) is applied, the surface buckles into wrinkles, showing some characteristic stripe patterns. The average wrinkle wavelength (λ) is typically $\lambda \approx 34 \pm 7$ μm at $s = 0.1$, suggesting that the surface layer with the porous film having an effective thickness D acts as a hard film with rigidity roughly close to isotropic bending rigidity on a macroscopic length scale. In this case of Fig. 4.11b, the wavelength shows a simple geometrical relationship given by $\lambda = n(3^{1/2}/2)[\Lambda(1 - s)]$, where n is an integer and is typically equal to 4 and Λ(~11 μm) is the periodicity of holes in the porous film. It should be noted that wrinkle crests and grooves appear at the middle of beams of the porous film, which are assumed to be the parts with the least bending rigidity. The wrinkle depth, $2A$, increases with s in an inverse-quadratic manner (Fig. 4.12a), which is typical of a wrinkling system.

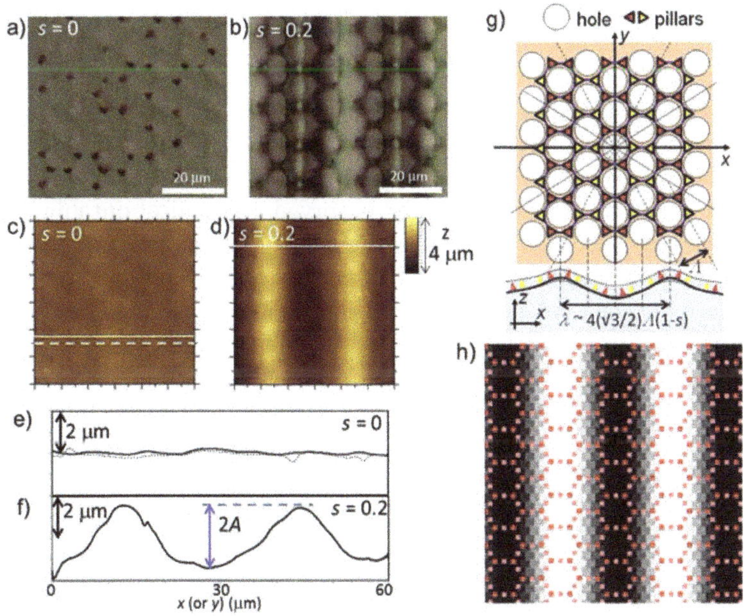

Figure 4.11 Wrinkling of the porous-film-embedded elastomer surface. (a, b) Optical microscopy images. (a) Image taken before strain application. (c, d) Atomic force microscopy (AFM) images (60^2 μm^2), and (e, f) corresponding AFM profiles along the lines indicated in (c) and (d), respectively. (c) Image taken after 100 compression–decompression cycles, showing neither clear damage nor undulations. (g) Schematic for the wrinkled state and the hexagonal regularity of the embedded porous film. (h) Profile obtained from a simple model simulation on a triangular lattice. At the red and pink points, which correspond to pillars and beams, respectively, the membrane has 10 and 5 times larger effective thickness than at the other points. Reproduced from Ref. [66] with permission from The Royal Society Chemistry.

Next, we demonstrate the mechanical durability of the present shape-tunable microwrinkles. Even after repeated compression–decompression cycles (~100) between s = 0 and 0.2, no clear difference is observed in the dependence of the wrinkle depth (2A) on strain (Fig. 4.12a). Moreover, the top surface shows neither cracks nor delamination. In contrast, PDMS simply coated with a polystyrene polymer (PSt) film (PSt/PDMS) easily starts showing delaminated crests and cracks perpendicular to the groove direction at s values around 0.1–0.15 (Fig. 4.12b). Although the present porous film appears to be partially delaminated from the PDMS matrix (Fig. 4.12a, inset) and to be plastically deformed, these cause no critical failure of the wrinkle structure. Another test involving mechanical scratching demonstrates that the present system has improved tribological durability (Fig. 4.12c,d). The PSt/PDMS system is easily destroyed by the present scratching condition. In contrast, the porous-film-embedded surface shows neither scars nor cracks after scratching. The observed mechanical durability is due to the interpenetrated structure between the gauche framework of the porous film and the PDMS matrix; this structure reinforces the effective hard layer for wrinkling and enlarges the material interfacial area, thereby preventing fatal delamination.

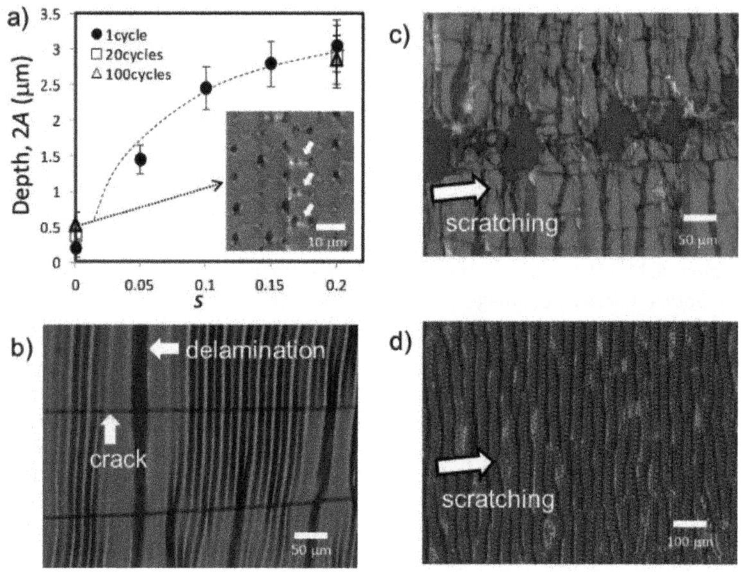

Figure 4.12 Mechanical durability. (a) Repeatable shape (depth) tunability of wrinkles formed on porous-film-embedded PDMS. (Inset) An optical microscopy image at $s = 0$ after 100 compression–decompression cycles for between $s = 0$ and $s = 0.2$. The positions indicated by white arrows correspond to the beams of the porous structure, at which a crest was located, and appear brighter than those of other beams. These parts are assumed to be slightly delaminated from PDMS inside the sample because of the repeated application of high strain. (b) PSt/PDMS under a strain of $s = 0.15$ shows wrinkles as well as delaminated crests and cracks, indicating low durability against high compressive strain. (c) PS/PDMS under a strain of $s = 0.15$ after a scratching test and showing critical failures. (d) Porous-film-embedded PDMS under a strain of $s = 0.15$ after a scratching test and showing original wrinkles without mechanical failure. Reproduced from Ref. [66] with permission from The Royal Society Chemistry.

Especially, the crack-resistant property on wrinkling, which is a major merit of using the present system, can be attributed to the textile-net-like flexibility of the porous film. The hexagons elongated in the y direction upon wrinkling (Fig. 4.11b) suggest that the tensile strain in the direction perpendicular to that of compression may be accommodated by rotating the beams with respect to the z axis. As a result, the actual tensile stress applied along the beams should be smaller than that found in the plane film under an identical external strain (or a scratching condition), improving crack resistance. This mechanical property is similar to that of the textile-net (see also Section 4.2.7). Moreover, even if some beams crack, the cracked parts are difficult to propagate because of discreteness of the porous structure, preventing critical failure of the wrinkled structure. In the next section, investigation of the friction on the porous-film-embedded wrinkles will be reviewed.

4.2.6 Friction with a Large Indenter on Porous-Film-Embedded Wrinkles

Here, the friction on porous-film-embedded wrinkles with the wavelength $\lambda \approx 34$ μm shown in Fig. 4.11 using an indenter ball with a large radius ($R = 5.0$ mm) is briefly reviewed; $R >> \lambda$ [67].

Figure 4.13 shows typical friction profiles on bare PDMS (BP), and flat (F) and wrinkled (W) porous-film-embedded surfaces. Note that all sample surfaces exhibited stick-slip-type oscillation of the friction force during sliding. The periodicities do not correspond to that of wrinkles, even on a W surface. The reason for the stick-slip phenomena originates from the sticky PDMS that covers most of the surface in contact. This is different from the previous profile found on the polyimide wrinkles.

Independent of the stick-slip phenomena, the average friction force, F_{av}, changes with respect to the sample surface type (Fig. 4.14). Specifically, the friction force with wrinkles was 25% less than that on the flat surface, which in turn was 35% smaller than bare PDMS (i.e., BP > F > W, Fig. 4.14). Moreover, although the friction force on the bare PDMS surface exhibited stick-slip oscillation with a large amplitude and Schallamach waves [55] in the early stages of steady sliding, the amplitude by which the friction force oscillated in stick-slip was relatively small in the case of the porous-film-embedded surfaces (F and W). Since groove structures were still evident, as indicated by the solid arrows in Fig. 4.15 (W-sliding), the grooves were clearly not in contact with the indenter during sliding.

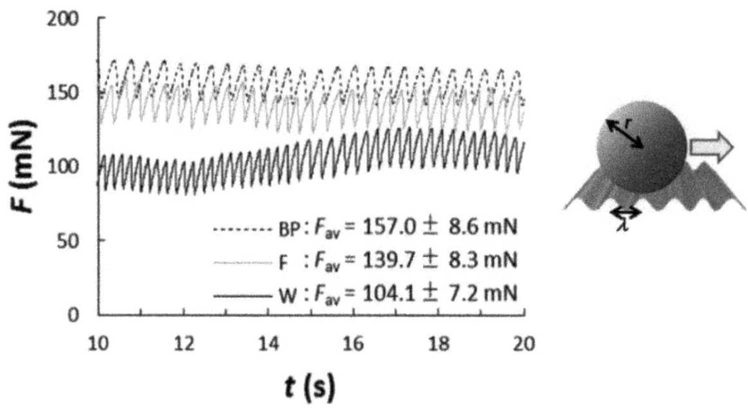

Figure 4.13 Friction profiles of steady oscillation states (10–20 s) under the load (P = 98.0 mN). Broken, gray, and solid lines in (a–d) are BP, F, and W surfaces, respectively. Adapted by permission from Springer Nature Customer Service Centre GmbH: Springer Nature, *Tribology Letters*, Ref. [67], Copyright (2015).

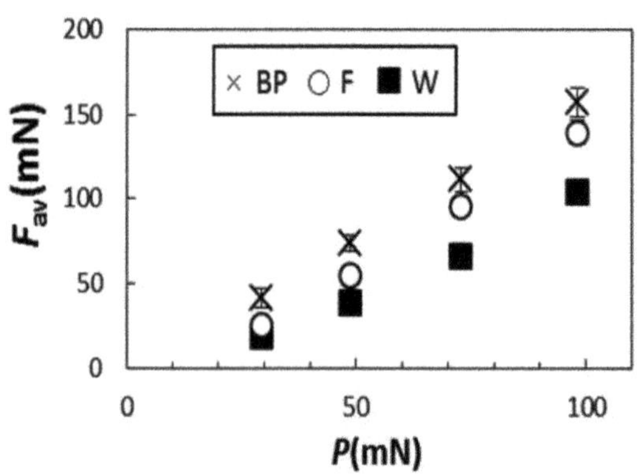

Figure 4.14 Average friction force F_{av} vs. normal load P. Adapted by permission from Springer Nature Customer Service Centre GmbH: Springer Nature, *Tribology Letters*, Ref. [67], Copyright (2015).

Generally, a linear relationship between friction and the real contact area ($F \propto \sigma A_r$, where σ is the shear stress to the rupture of adhesion and A_r is the real contact area) has been proposed for the solid friction systems, which is well known as Bowden–Tabor's law [52]. On the basis of this framework, the present qualitative result can be explained as follows. First, the combination of materials at the frictional interface in all experiments (BP, F, and W) can be considered equivalent in terms of the surface chemistry, and thus σ is constant. It therefore follows that a decrease in real contact area due to the increased modulus of the embedded-film rubber surface is the direct cause of the reduction in friction observed.

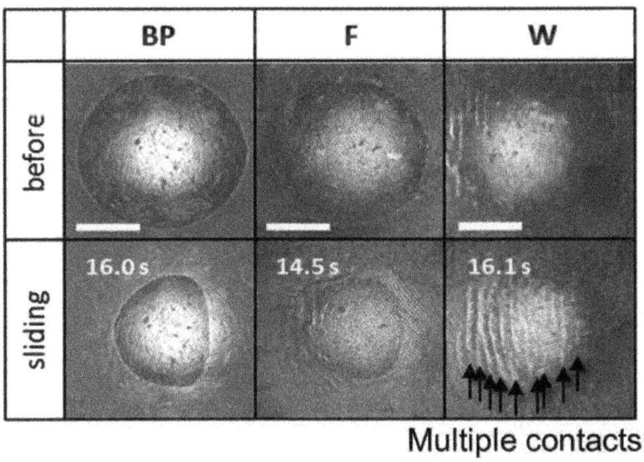

Figure 4.15 Optical micrographs of the friction interfaces before (left) and during (right) sliding under the optical microscope. The direction of sliding was from right to left (i.e., perpendicular to the grooves, which are indicated by solid arrows at the friction interfaces), with the position of the indenter fixed. Scale bars = 400 μm. Adapted by permission from Springer Nature Customer Service Centre GmbH: Springer Nature, *Tribology Letters*, Ref. [67], Copyright (2015).

4.2.7 Preparation of Textile-Sheet-Embedded Wrinkles

Another version of the shape-tunable wrinkles with mechanical durability toward tribological uses is shown here [68]. Instead of using the porous film shown in the previous two sections, a textile sheet is used as the surface skin layer for buckling to induce wrinkles. The textile sheet is more of a general material than the self-organized porous film and, thus, is commercially available in large sizes at a low cost. This section briefly describes the fabrication method of wrinkles formed on a textile-embedded elastomer.

The fabrication method is summarized in Fig. 4.16a. A biaxial textile sheet, that is a plain weave fabric made of nylon-66 fibre (yarn) with a diameter of d = 35 μm was used. The periodicity of the warp and weft p is approximately 80 μm (Fig. 4.16a, inset). This was used as a hard porous sheet to be embedded within a layer of a PDMS elastomer. A portion of the textile sheet was placed on a flat glass plate, and the liquid precursor of the PDMS was poured onto the sheet. A 5-mm-thick elastomeric silicone rubber base piece was steadily pressed onto the sample to squeeze out the excessive liquid precursor and smoothen the textile sheet out in a reproducible manner, after which it was cured at 80°C for 3 h. The cured sample was peeled from the glass substrate. The sample was uniaxially compressed in the lateral direction using a small vice to induce wrinkles on the surface (Fig. 4.16a). The applied compressive strain was changed to tune the depth and shape of the wrinkles (Fig. 4.17).

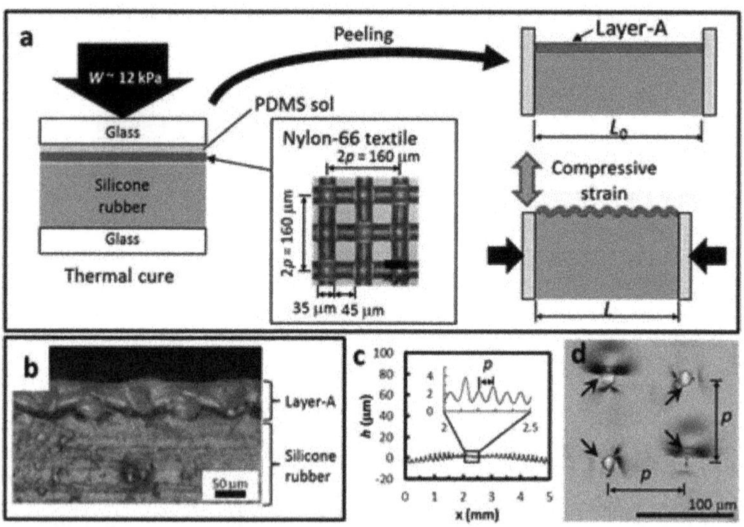

Figure 4.16 Textile-embedded elastomer sample. (a) Schematic illustration for sample fabrication. (b) Cross-sectional image and (c) surface profile of sample surface at s = 0. (d) Optical image of sample surface at s = 0. Note that the skin of the PDMS on the intersection point is partially peeled by a diameter of 15–25 μm, indicated by arrows. Reproduced from Ref. [68] with permission from The Royal Society Chemistry.

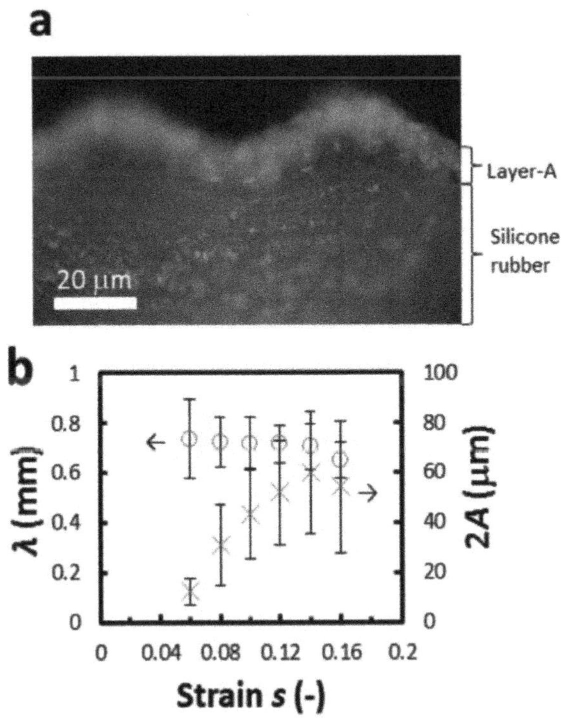

Figure 4.17 Surface of textile-embedded elastomer under uniaxial compressive strain. (a) Cross-sectional image of the wrinkled sample and (b) plots of wavelength λ (circles) and amplitude $2A$ (crosses) with respect to applied strain s. Reproduced from Ref. [68] with permission from The Royal Society Chemistry.

4.2.8 Friction with a Large Indenter on Textile-Sheet-Embedded Wrinkles

Here, the friction on textile-embedded wrinkles with the wavelength $\lambda \approx 0.7$ mm shown in Fig. 4.18 using an indenter ball with a large radius ($R = 5.0$ mm) is briefly reviewed; $R > \lambda$ [68].

The average friction force, F_{av}, on the flat state shows high values that are similar to those of rubbers (Fig. 4.18). In contrast, F_{av} on wrinkles under compressive strain of 16% decreases significantly by a factor of ~1/12 compared to that on an unwrinkled surface. Additionally, the averaged friction force for the sliding in the direction parallel to the wrinkle grooves also shows low values (data not shown).

To reveal the mechanism for this significant reduction of friction on the textile-embedded wrinkles, the contact states were evaluated by fluorescent-dye-stamping method. The lower panels on Fig. 4.18 show the representative fluorescent images of the traces of contact on unwrinkled and wrinkled surfaces. On the unwrinkled surface, the apparent contact area appears as a single circular region expected from a simple Hertzian contact. In striking contrast, the contact region on the wrinkled surface is limited to the alternatively protruding intersection points on top of the two neighboring wrinkle crests, the exposed warp nylon-66 fiber, and the small surrounding area covered by PDMS, which appears as ring-like brighter edge in the fluorescent images. This significant reduction of the apparent contact area may be probably the main reason for the greatly reduced friction, assuming that the true contact area in the Bowden–Tabor's law is qualitatively related to the apparent one and that the shear strengths for both high and low friction states are in the same order of magnitude.

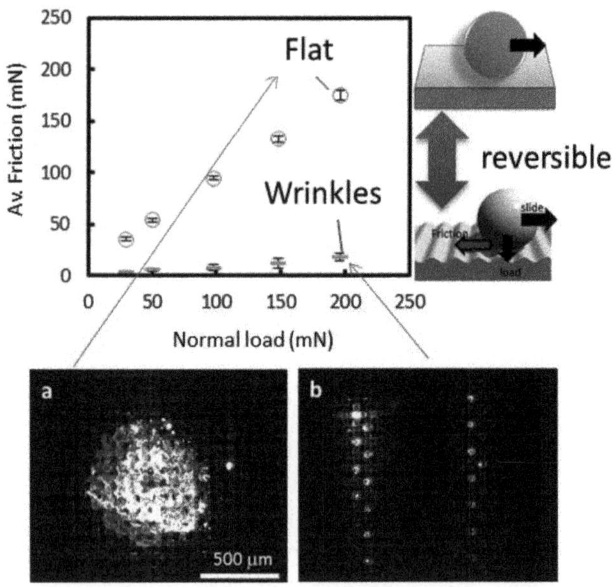

Figure 4.18 Averaged friction forces with respect to normal load on the surface of a textile-embedded elastomer with flat or wrinkled state. The lower panels show typical fluorescent images for estimation of the contact state during friction on flat and wrinkled surfaces. Reproduced from Ref. [68] with permission from The Royal Society Chemistry.

4.3 Summary

In this chapter, some recent progress of the studies on frictional properties of various shape-tunable wrinkles has been reviewed with a brief introduction of the shape-tunable wrinkles. In Table 4.1, the present results of frictional properties are crudely summarized in terms of the relative size of the wrinkle periodicity and the radius of the indenter. It is noted that the wrinkled states can either increase and decrease friction, depending on the relative length scale. This is a very important message for designing a frictional system with the ability to switch the friction using wrinkles toward a certain application. This result also inspires that the effect of the wrinkles may be greatly affected by the hardness of the indenter, sliding velocity, and lubricants, which have not been studied yet. Moreover, the friction on wrinkles at a high load range may also show abnormal dependency because of the deformation of the structures, which also remains to be studied. Nevertheless, the present system of shape-tunable wrinkles is believed to be encouraging as a unique switchable surface on which the friction coefficient is changeable and may lead to further creation of new tribological soft interfaces, some of which may have already been utilized in biological systems and not discovered yet.

Table 4.1 Preliminary summary of frictional properties of various shape-tunable wrinkles

Hard skin layer	Substrate	Typical wrinkle periodicity (μm)	Radius of frictional hard indenter (mm)	Lower friction on?
Polyimide	PDMS	500-600	1.0	Flat
Polyimide	PDMS	500-600	10.0	Wrinkles
Porous film (Honeycomb film)	PDMS	~30	5.0	Wrinkles
Textile sheet	PDMS	~600	5.0	Wrinkles

Acknowledgments

This work was partly supported by the New Energy and Industrial Technology Development Organization (NEDO) of Japan under the Industrial Technology Research Grant Program in 2008 and JSPS KAKENHI Grant Numbers JP24120003 and JP24120001 in Scientific Research on Innovative Areas "Innovative Materials Engineering on Biological Diversity." The author thanks H. Monobe, Y. Shimizu, H. Yokoyama, M. Shimomura, Y. Hirai, N. Uchida, K. Teraoka, and T. Yamaguchi for their helpful comments and experimental supports.

References

1. Nosonovsky, M. and Bhushan, B. (2008). Friction, Multiscale Dissipative Mechanisms and Hierarchical Surfaces: Superhydrophobicity, and Biomimetics (Springer, Berlin, Heidelberg).
2. Autumn, K., Liang, Y. A., Hsieh, S. T., Zesch, W., Chan, W. P., Kenny, T. W., Fearing, R. and Full, R. J. (2000). Adhesive force of a single gecko foot-hair, Nature, 405, pp. 681–685.
3. Federle, W., Barnes, W. J. P., Baumgartner, W., Drechsler P. and Smith, J. M. (2006). Wet but not slippery: boundary friction in tree frog adhesive toe pads, J. R. Soc. Interface, 3, pp. 689–697.
4. Kier, W. M. and Smith, A. M. (2002). The structure and adhesive mechanism of octopus suckers, Integr. Comp. Biol., 42, pp. 1146–1153.
5. Voigt, D., Schweikart, A., Fery, A. and Gorb, S. (2012). Leaf beetle attachment on wrinkles: isotropic friction on anisotropic surfaces, J. Exp. Biol., 215, pp. 1975–1982.
6. Prüm, B., Bohn, H. F., Seidel, R., Rubach, S. and Speck, T. (2013). Plant surfaces with cuticular folds and their replicas: influence of microstructuring and surface chemistry on the attachment of a leaf beetle, Acta Biomater., 9, pp. 6360–6368.
7. Bohn, H. F. and Federle, W. (2004). Insect aquaplaning: nepenthes pitcher plants capture prey with the peristome, a fully wettable water-lubricated anisotropic surface, Proc. Natl. Acad. Sci. U. S. A., 101, pp. 14138–14143.
8. Nonomura, Y., Miura, T., Miyashita, T., Asao, Y., Shirado, H., Makino, Y. and Maeno, T. (2012). How to identify water from thickener aqueous solutions by touch, J. R. Soc. Interface, 9, pp. 1216–1223.

9. Adams, M. J., Johnson, S. A., Lefèvre, P., Lévesque, V., Hayward, V., André, T. and Thonnard, J.-L. (2013). Finger pad friction and its role in grip and touch, J. R. Soc. Interface, 10, p. 20120467.
10. Skedung, L., Arvidsson, M., Chung, J. Y., Stafford, C. M., Berglund, B. and Rutland, M. W. (2013). Feeling small: exploring the tactile perception limits, Sci. Rep., 3, p. 2617.
11. Grosch, K. A. (1963). The relation between the friction and visco-elastic properties of rubber, Proc. R. Soc. London, Ser. A, 274, pp. 21–39.
12. Persson, B. N. J. (2001). Theory of rubber friction and contact mechanics, J. Chem. Phys., 115, pp. 3840–3861.
13. Tominaga, T., Kurokawa, T., Furukawa, H., Osada, Y. and Gong, J. P. (2008). Friction of a soft hydrogel on rough solid substrates, Soft Matter, 4, pp. 1645–1652.
14. Popov, V. L. and Filippov, A. É. (2010). Force of friction between fractal rough surface and elastomer, Tech. Phys. Lett., 36, pp. 525–527.
15. Kürschner, S. and Popov, V. L. (2013). Penetration of self-affine fractal rough rigid bodies into a model elastomer having a linear viscous rheology, Phys. Rev. E, 87, p. 042802.
16. Pohrt, R., Popov, V. L. and Filippov A. E. (2012). Normal contact stiffness of elastic solids with fractal rough surfaces for one- and three-dimensional systems, Phys. Rev. E, 86, p. 026710.
17. Popov, V. L., Voll, L., Li, Q., Chai, Y. S. and Popov, M. (2013). Contact stiffness of randomly rough surfaces, Sci. Rep., 4, p. 3750.
18. Sneddon, I. N. (1965). The relation between load and penetration in the axisymmetric boussinesq problem for a punch of arbitrary profile, Int. J. Eng. Sci., 3, pp. 47–57.
19. Johnson, K. L., Kendall, K. and Roberts, D. (1971). Surface energy and the contact of elastic solids, Proc. R. Soc. London, Ser. A, 324, pp. 301–313.
20. Johnson, K. L., Greenwood, J. A. and Higginson, J. G. (1985). The contact of elastic regular wavy surfaces, Int. J. Mech. Sci., 27, pp. 383–396.
21. Johnson, K. L. (1997). Adhesion and friction between a smooth elastic spherical asperity and a plane surface, Proc. R. Soc. London, Ser. A, 453, pp. 163–179.
22. Guduru, P. R. (2007). Detachment of a rigid solid from an elastic wavy surface: theory, J. Mech. Phys. Solids, 55, pp. 445–472.
23. Waters, J. F., Lee, S. and Guduru, P. R. (2009). Mechanics of axisymmetric wavy surface adhesion: JKR-DMT transition solution, Int. J. Solids Struct., 46, pp. 1033–1042.

24. Lin, P.-C., Vajpayee, S., Jagota, A., Huid, C.-Y. and Yang, S. (2008). Mechanically tunable dry adhesive from wrinkled elastomers, Soft Matter, 4, pp. 1830–1835.

25. Jin, C., Khare, K., Vajpayee, S., Yang, S., Jagota, A. and Hu, C.-Y. (2011). Adhesive contact between a rippled elastic surface and a rigid spherical indenter: from partial to full contact, Soft Matter, 7, pp. 10728–10736.

26. Kundu, S., Davis, C. S., Long, T., Sharma, R. and Crosby, A. J. (2011). Adhesion of nonplanar wrinkled surfaces, J. Polym. Sci., Part B: Polym. Phys., 49, pp. 179–185.

27. Geim, A. K., Dubonos, S. V., Grigorieva, I. V., Novoselov, K. S., Zhukov, A. A. and Shapoval, S. Y. (2003). Microfabricated adhesive mmicking gecko foot-hair, Nat. Mater., 2, pp. 461–463.

28. Chateauminois, A. and Fretigny, C. (2008). Local friction at a sliding interface between an elastomer and a rigid spherical probe, Eur. Phys. J. E: Soft Matter Biol. Phys., 27, pp. 221–227.

29. Rand, C. J. and Crosby, A. J. (2009). Friction of soft elastomeric wrinkled surfaces, J. Appl. Phys., 106, p. 064913.

30. Wu-Bavouzet, F., Cayer-Barrioz, J., Le Bot, A., Brochard-Wyart, F. and Buguin, A. (2010). Effect of surface pattern on the adhesive friction of elastomers, Phys. Rev. E, 82, p. 031806.

31. Wandersman, E., Candelier, R., Debrégeas, G. and Prevost, A. (2011). Texture-induced modulations of friction force: the fingerprint effect, Phys. Rev. Lett., 107, p. 164301.

32. Kim, S. J., Yoon, J.-I., Moon, M.-W. and Lee, K.-R. (2012). Frictional behavior on wrinkle pattern of diamond-like carbon films on soft polymer, Diamond Relat. Mater., 23, pp. 61–65.

33. Audry, M. C., Fretigny, C., Chateauminois, A., Teissere, J. and Barthel, E. (2012). Slip dynamics at a patterned rubber/glass interface during stick-slip motions, Eur. Phys. J. E: Soft Matter Biol. Phys., 35, pp. 83–90.

34. Ramakrishna, S. N., Espinosa-Marzal, R. M., Naik, V. V., Nalam, P. C. and Spencer, N. D. (2013). Adhesion and friction properties of polymer brushes on rough surfaces: a gradient approach, Langmuir, 29, pp. 15251–15259.

35. Waters, J. F. and Guduru, P. R. A (2011). Mechanism for enhanced static sliding resistance owing to surface waviness, Proc. R. Soc. London, Ser. A, 467, pp. 2209–2223.

36. Bowden, N., Brittain, S., Evans, A. G., Hutchinson J. W. and Whitesides, G. M. (1998). Spontaneous formation of ordered structures in thin

films of metals supported on an elastomeric polymer, Nature, 393, pp. 146–149.

37. Bowden, N., Huck, W. T. S., Paul, K. E. and Whitesides, G. M. (1999). The controlled formation of ordered, sinusoidal structures by plasma oxidation of an elastomeric polymer, Appl. Phys. Lett., 75, pp. 2557–2559.
38. Brau, F., Vandeparre, H., Sabbah, A., Poulard, C., Boudaoud, A. and Damman, P. (2011). Multiple-length-scale elastic instability mimics parametric resonance of nonlinear oscillators, Nat. Phys., 7, pp. 56–60.
39. Yang, S., Khare, K. and Lin, P.-C. (2010). Harnessing surface wrinkle patterns in soft matter, Adv. Funct. Mater., 20, pp. 2550–2564.
40. Schallamach, A. (1971). How does rubber slide? Wear, 17, pp. 301–312.
41. Rand, C. J. and Crosby, A. J. (2006). Insight into the teriodicity of Schallamach waves in soft material friction, Appl. Phys. Lett., 89, p. 261907.
42. Ohzono, T., Suzuki, K., Yamaguchi, T. and Fukuda, N. (2013). Tunable optical diffuser based on deformable wrinkles, Adv. Opt. Mater., 1, pp. 374–380.
43. Ohzono, T. and Shimomura, M. (2004). Ordering of microwrinkle patterns by compressive strain, Phys. Rev. B, 69, p. 132202.
44. Ohzono, T., Watanabe. H., Vendame, R., Kamaga, T., Ishihara, T., Kunitake, T. and Shimomura, M. (2007). Spatial forcing of self-organized microwrinkles by periodic nanopatterns, Adv. Mater., 19, pp. 3229–3232.
45. Ohzono, T. (2009). Control of cooperative switching of microwrinkle orientations by nanopatterns, Chaos, 19, p. 033104.
46. Ohzono, T. and Monobe, H. (2012). Microwrinkles: shape-tunability and applications, J. Colloid Interface Sci., 369, pp. 1–8.
47. Ohzono, T., Monobe, H., Yamaguchi, R., Shimizu, Y. and Yokoyama, H. (2009). Dynamics of surface memory effect in liquid crystal alignment on reconfigurable microwrinkles, Appl. Phys. Lett., 95, p. 014101.
48. Ohzono, T. and Fukuda, J. (2012). Zigzag line defects and manipulation of colloids in a nematic liquid crystal in microwrinkle grooves, Nat. Commun., 3, p. 701.
49. Wang, C., Hwang, D., Yu, Z., Takei, K., Park, J., Chen, T., Ma, B. and Javey, A. (2013). User-interactive electronic-skin for instantaneous pressure visualization, Nat. Mater., 12, pp. 899–904.
50. Sharp, J. S. and Jones, R. A. L. (2002). Micro-buckling as a route towards surface patterning, Adv. Mater., 14, pp. 799–802.

51. Vendamme, R., Onoue, S., Nakao, A. and Kunitake, T. (2006). Robust free-standing nanomembranes of organic/inorganic interpenetrating networks, Nat. Mater., 5, pp. 494–501.

52. Bowden, F. P. and Tabor, D. (1950). The Friction and Lubrication of Solids (Oxford University Press, New York).

53. Rabinowicz, E. (1995). Friction and Wear of Materials, 2nd ed. (John Wiley & Sons, New York).

54. Popov, V. L. (2010). Contact Mechanics and Friction: Physical Principles and Applications (Springer–Verlag and Heidelberg GmbH Co. KG, Berlin).

55. Schallmach, A. (1958). Friction and abrasion of rubber, Wear, 1, pp. 384–417.

56. Persson, B. N. J. (1998). On the theory of rubber friction, Surf. Sci., 401, pp. 445–454.

57. Jagota, A. and Hui, C.-Y. (2011). Adhesion, friction, and compliance of bio-mimetic and bio-inspired structured interfaces, Mater. Sci. Eng. Rep., 72, pp. 253–292.

58. Brörmann, K., Barel, I., Urbakh, M. and Bennewitz, R. (2013). Friction on a microstructured elastomer surface, Tribol. Lett., 50, pp. 3–15.

59. Bhushan, B. (2009). Biomimetics: lessons from nature - an overview, Phil. Trans. R. Soc. A, 367, pp. 1445–1486.

60. Akiyama, H. and Yoshida, M. (2012). Photochemically reversible liquefaction and solidification of single compounds based on a sugar alcohol scaffold with multi azo-arms, Adv. Mater., 24, pp. 2353–2356.

61. Suzuki, K., Hirai, Y. and Ohzono, T. (2014). Oscillating friction on shape-tunable wrinkles, ACS Appl. Mater. Interfaces, 6, pp. 10121–10131.

62. Kaneko, D., Oshikawa, M., Yamaguchi, T., Gong, J. P. and Doi, M. (2007). Friction coefficient between rubber and solid substrate –effect of rubber thickness–, J. Phys. Soc. Jpn., 76, p. 043601.

63. Adams, M. J., Briscoe, B. J. and Wee, T. K. (1990). The differential friction effect of keratin fibres, J. Phys. D: Appl. Phys., 23, pp. 406–414.

64. Suzuki, K. and Ohzono, T. (2016). Friction on non-uniform wrinkled surface, Kobunshi Ronbunshu, 6, pp. 514–519 (in Japanese).

65. Hirai, Y., Yabu, H., Matsuo, Y., Ijiro, K. and Shimomura, M. (2010). Preparation of self-organized porous polymer masks for Si dry etching, J. Mater. Chem., 20, pp. 10804–10808.

66. Ohzono, T., Hirai, Y., Suzuki, K., Shimomura, M. and Uchida, N. (2014). Reinforced shape-tunable microwrinkles formed on a porous-film-embedded elastomer surface, Soft Matter, 10, pp. 7165–7169.

67. Suzuki, K., Hirai, Y. and Ohzono, T. (2015). Tunable friction through microwrinkle formation on a reinforced rubber surface, Tribol. Lett., 60, p. 24.

68. Suzuki, K. and Ohzono, T. (2016). Wrinkles on a textile-embedded elastomer surface with highly variable friction, Soft Matter, 12, p. 6176.

Chapter 5

Self-Lubricating Gels: SLUGs

Atsushi Hozumi, Chihiro Urata, Tomoya Sato, Liming Wang, and Matt W. England

National Institute of Advanced Industrial Science and Technology (AIST) 2266-98 Anagahora, Shimo-shidami, Nagoya 463-8560, Japan

a.hozumi@aist.go.jp

Self-lubricating gels (SLUGs), artificially mimicking antifouling/self-cleaning functionalities of a slug's body, were successfully prepared via a simple cross-linking of polydimethylsiloxane (PDMS) in the presence of several organic liquids. Due to the syneresis of the organogels, a thin and uniform liquid layer was formed on the topmost SLUG surfaces under the proper conditions. When the appropriate organic liquids (reactive/unreactive) were selected, the SLUG surfaces showed unusual dynamic dewetting behavior toward various viscous emulsions, spontaneous formation/regeneration of superhydrophobicity, and thermoresponsive anti-icing/snow properties.

Industrial Biomimetics

Edited by Akihiro Miyauchi and Masatsugu Shimomura

ISBN 978-981-4800-07-5 (Hardcover), 978-0-429-05883-7 (eBook)

www.jennystanford.com

5.1 Introduction

The formation of functional coatings that exhibit outstanding surface properties, such as superwettability and low friction/adhesion, has generally relied on a combination of surface structures and low-surface-energy materials (typically long-chain perfluorinated compounds) [1–3], and although such man-made surfaces often display excellent surface properties initially, once they are physically/chemically damaged their surface functionalities will quickly, and often irreversibly, worsen. Such artificial surfaces typically show no regeneration of their functionalities after being damaged, and this serious shortcoming significantly restricts any potential applications. To overcome this, and attain long-lasting surface functionality, several potential approaches have so far been reported [4–7]. In contrast, many living organisms can sustain their surface properties through continuous secretion of waxes or mucus [8], and therefore, to realize long-lasting surface functionalities similar to those observed in biological surfaces, we have focused on the phenomenon of "syneresis," which is the release of liquids from inner gel matrices onto their outer surfaces.

In this chapter, we introduce our recently developed novel organogels, which we have named "self-lubricating gels," (SLUGs), showing unusual and unique dynamic surface functionalities such as anti(bio)fouling, anti-icing/snow, and self-healing superhydrophobic properties similar or even superior to biological surfaces [9].

5.2 Sample Preparation

Typically, SLUGs can be prepared by finely tuning the ratios of organic liquid phases (mobile, reactive/unreactive, and non-cross-linked components in gel matrices) to a solid phase (a three dimensionally cross-linked gel network) [9]. From among the various gel matrices available, we selected cross-linked polydimethylsiloxane (PDMS) as a host gel matrix because it is transparent, flexible, and elastic and has a high affinity for several liquids with low vapor pressures. Such physical properties were expected to prevent cracking, even after the reduction in volume caused by syneresis.

In this study, the commercially available Sylgard® 184 Silicone Elastomer Kit was used to prepare PDMS gel matrices. The organic liquid phase(s) (vol%, α ranging from 0 to 3000) were mixed with 1 g of the base unit and 0.1 mL of a curing agent to form a precursor liquid. This precursor solution was then poured into a Teflon® container and heated in an oven maintained at 100°C for 3 h to promote gelation.

The organic liquids used in this study, which show different molecular interactions with the solid phase, are summarized in Table 5.1 (with water as a control). We have particularly focused on the changes in miscibility of the organic liquid phase before and after gelation of the solid phase. These two phases are miscible with each other in the precursor state, as the free energy of mixing ($\Delta G = \Delta H - T\Delta S$) is negative ($\Delta G < 0$) [10]. However, cross-linking of the gel during/after gelation results in an increase in ΔG, which can lead to demixing (if ΔG becomes greater than 0). Thus, the organic liquid phase, which was once homogeneously mixed ($\Delta G < 0$), can spontaneously leach out from the organogel surface after gelation. This leaching is driven by a change in the enthalpy of mixing ΔH, which in this system is given by $\Delta H = V(\delta_1 - \delta_2)^2\varphi_1\varphi_2$ (where V is the volume of the mixture, δ_1 and δ_2 are the solubility parameters of gel precursor and organic liquid, and φ_1 and φ_2 are the volume fractions of gel precursor and organic liquid, respectively). Specifically, the solubility parameters $(\delta_1 - \delta_2)^2$ are altered by the occurring cross-linking reactions. The stretching of polymer chains during cross-linking, which leads to a reduction in entropy, may also contribute to changes in the free energy of mixing, but the majority of this change comes from enthalpic contributions.

From our preliminary experiments, we classified the 17 liquids used into three categories, Class (I), Class (II), and Class (III), on the basis of miscibility and syneresis-inducing properties. Class (I) liquids ($\Delta G_{precursor} < 0 \Rightarrow \Delta G_{gel} < 0$) were miscible with PDMS before and after gelation, giving no syneresis; Class (II) liquids ($\Delta G_{precursor} < 0 \Rightarrow \Delta G_{gel} > 0$) were miscible with PDMS before gelation but showed syneresis behavior because of low compatibility with the PDMS matrices; and Class (III) liquids ($\Delta G_{precursor} > 0$) were immiscible with the PDMS precursor but miscible with Class (I) and/or (II) organic liquids.

Table 5.1 Organoliquids used in this study (water used as a control)

No.	Liquids	
1	*n*-Hexane	Class (I)
2	*n*-Octane	
3	Toluene	
4	*n*-Decane	
5	Isocetane	
6	*n*-Dodecane	
7	Polydimethylsiloxane (PDMS)	
8	*n*-Tetradecane	
9	Polymethylphenylsiloxane (PMPS, AR20)	
10	*n*-Hexadecane	Class (II)
11	PMPS (CR100)	
12	Phenylmethylsiloxane-dimethylsiloxane copolymer (PMPS-PDMS, AS100)	
13	PMPS (TSF431)	
14	PMPS (TSF437)	Class (III)
15	Octadecyltrichororosilane (ODS)	
16	PMPS-PDMS (AP100)	
17	Water	

We first prepared organogels using Class (I) or (II) liquids and investigated the optimum conditions to achieve transparent organogels and the critical points for gelation. Transparent organogels could be obtained when the value of α was lower than 1200, independent of the organic liquids. Judging from scanning electron microscope (SEM) images (not shown) the PDMS surfaces were confirmed to be fairly smooth, even after extraction of the organic phases. No marked differences were found between the surface morphologies of the organic-phase-extracted PDMS and the original PDMS surfaces. We thus concluded that the PDMS matrix and organic liquids were homogeneously mixed on a microscopic level. However, when α exceeded 1200 the gel was only lightly cross-linked and did not form a gel. Above this critical point for gelation, the gel networks became discontinuous due to the decrease in the PDMS available for polymer network formation.

For example, after gelation, no syneresis was observed on an organogel prepared with AR20—a Class (I) liquid, no. 9 in Table 5.1, α = 300—because of the high compatibility between the solid and liquid phases (we hereafter refer to this nonsyneretic organogel as NSG). In contrast, the organogel prepared with TFS431—a Class (II) liquid, no. 13 in Table 5.1, α = 300—exhibited syneresis (we hereafter refer to this syneretic organogel as a SLUG). As shown in Fig. 5.1, in the case of SLUGs, syneresis was also observed on newly generated surfaces, formed by cutting the SLUGs. The fractured and opaque cross-section of the newly formed surface became fairly smooth after 1 h due to syneresis, leading to an increase in the transparency of the final sample. In contrast, the fractured and opaque cross-section of an NSG remained unchanged.

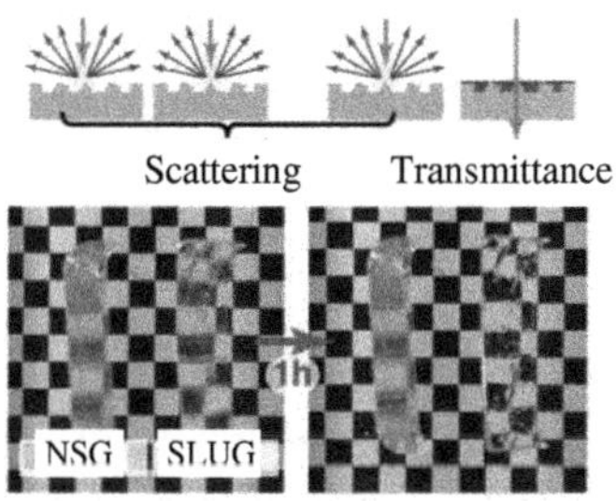

Figure 5.1 Changes in optical properties of the newly formed surfaces of NSG (AR20) and SLUG (TSF431), 1 h after the samples were cut (cross sections were deliberately roughened using a box cutter).

In addition, the samples prepared with Class (II) liquids (no. 11–13 in Table 5.1) exhibited long-lasting syneretic behavior compared to those prepared with *n*-hexadecane—a Class (II) liquid, no. 10 in Table 5.1. The weight of these three samples continued to gradually decrease, even after they were left in air for more than three months. The syneretic behavior could also be finely controlled by appropriate tuning of the organic liquid phases with the PDMS precursor by mixing a Class (I) liquid with a Class (II) liquid [9].

5.3 Applications of SLUGs

5.3.1 Liquid Repellency

We first investigated the water-repellent properties of the samples, that is, (1) the PDMS resin (control) (2) NSG (AR20); and the SLUG in the (3) initial and (4) equilibrium states (TSF431).

As shown in Fig. 5.2, water droplets would completely stick to the inclined PDMS resin surface, while they could slide off from the NSG and SLUG surfaces at differing velocities. The maximum sliding speed achieved for water was around 5 cm/s, which occurred after 6 h of syneresis. After this amount of time, the thickness of the liquid layer was estimated to be about 20 μm. In addition, even after removing the liquid layer by wiping, it was gradually reformed by syneresis, and the excellent dynamic dewettability was recovered after three days; this dewettability continued when our SLUGs were left in air for almost one year.

These differences in observed water sliding speeds can be explained in the following three ways. Firstly, in the case of the NSG surface the presence of both the mobile-organic-liquid phase and the PDMS network may have reduced interaction between the water and the PDMS network, resulting in superior water repellency, not observed on the initial PDMS resin. Secondly, the physicochemical parameters of the infused liquids, particularly viscosity, played a key role in determining the sliding speed. For example, as shown in Fig. 5.2, on the NSG surface (viscosity of AR20 = 20 mPa/s), the sliding speed of the water droplet was around three times faster than that on the initial SLUG surface (viscosity of TSF431 = 100 mPa/s). It has been reported that the velocity of water droplets on inclined slippery liquid-infused porous surfaces (SLIPS) markedly increased with decreasing viscosity of the liquids used to impregnate the surface [11]. However, the reason for the velocity's viscosity dependence has not yet been identified. Thirdly, after 6 h of syneresis, due to the formation of the continuous/sufficient thick syneresis liquid layer on the SLUG surface, sliding speed of the water droplet was about 1.7 times faster than that on the NSG surface. In this case, not only water but also viscous liquids (mayonnaise, liquid glue, ketchup, honey, and Worcester sauce) could be easily removed from the surface with negligible adhesion force. Conversely, on the NSG surface, all emulsions adhered to the surface and only Worcester sauce and honey flowed (but these left surface residues).

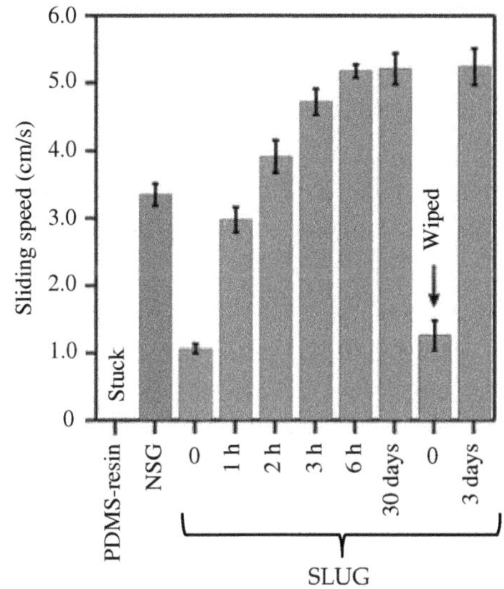

Figure 5.2 Average sliding speeds of water drops (50 μL) on the inclined (30°) PDMS-resin, NSG (AR20), and SLUG (TSF431).

5.3.2 Spontaneous Formation/Regeneration of Superhydrophobicity

Another advantage of our system is that insoluble organic liquids—Class (III) liquids, no. 14–17 in Table 5.1—can be incorporated into the gel matrices using mixed organic liquid phases. Using this system we could successfully fabricate surfaces that showed spontaneous superhydrophobicity and self-healing properties, even after mechanical damage. It is well known that reactive organosilanes, such as *n*-octadecyltrichlorosilane (ODS, $C_{18}CH_2CH_2SiCl_3$—a Class (III) liquid, no. 15 in Table 5.1) can self-assemble to form textured superhydrophobic surfaces after undergoing sol-gel reactions [12, 13].

The initially transparent ODS-infused SLUG became gradually opaque after 3 h of exposure to air (60% relative humidity) due to the formation of textured structures on the surface, as confirmed by SEM images (Fig. 5.3). The resulting surface displayed superhydrophobicity with a large static water contact angle (CA) of over 150°. Water droplets could move smoothly, without getting pinned on our superhydrophobic SLUG surface because of negligible CA hysteresis. In contrast, the inside of the ODS-infused SLUG remained transparent, indicating that the hydrolysis and polycondensation reactions of the ODS molecules only took place in the outermost regions, close to the surface. Our superhydrophobic SLUG surfaces also possessed excellent self-healing properties through the regeneration of microstructures, as demonstrated by the transparent regions of cut cross-sections, which became opaque and once again showed superhydrophobicity after exposure to air for 24 h. Even after the ODS-infused SLUG underwent severe mechanical damage from an electric drill its superhydrophobicity could be similarly regenerated by leaving the sample in air for 24 h (Fig. 5.4).

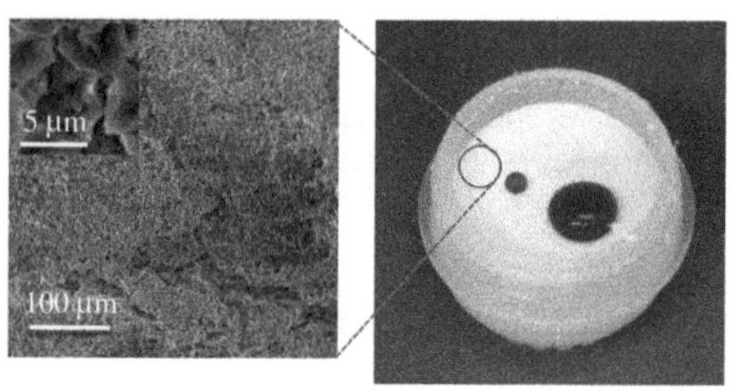

Figure 5.3 Water droplets on the ODS-infused SLUG after 3 h of exposure to air, and SEM images of the sample surface.

Figure 5.4 Self-healing properties of ODS-infused SLUG: (a) Surface microstructures were collapsed by an electric drill. (b) Water droplets on the damaged surface. (c) Regeneration of superhydrophobicity after leaving sample in air for 24 h.

5.3.3 Anti-icing/Snow Properties

We further demonstrated that we could control syneretic behavior by simply varying the temperature. To demonstrate this, we selected AR20 and TSF437—a Class (III) liquid, no. 14 in Table 5.1—as organic phases. For samples where the proportion of TSF437 was less than 80% of the total organic liquid phase (by volume), TSF437 was miscible with the PDMS precursors. After gelation, TSF437 was incorporated into the gel matrix, forming a transparent organogel. However, the syneretic behavior of these organogels was markedly dependent on the proportional volume of TSF437. In cases where TSF437 made up 40%–80% syneresis was observed at room temperature (RT) due to the change in ΔG caused by gelation ($\Delta G_{precursor} < 0 \Rightarrow \Delta G_{gel} > 0$). Thus, this mixture of organic phases behaved like a Class (II) organic phase. Unusually, syneresis of TSF437 was not observed at RT in samples below 30% TSF437, but when the samples were cooled to −15°C liquid leached out from the surface. This clearly indicates that our SLUGs show thermoresponsive properties (we hereafter refer to these thermoresponsive organogels as TR-SLUG). Furthermore, the syneresis liquid on the TR-SLUG surface was gradually reabsorbed back into the gel matrix after the samples were warmed back to RT over the course of 3 h. This unusual reversible thermoresponsive behavior was found to be repeatable for at least 10 cycles and enabled the same liquid to be reused many times, preventing undesirable loss of syneresis liquids.

Finally, we tested our TR-SLUG as a potential anti-icing/snow surface. As shown in Fig. 5.5, impregnation of organic liquids greatly reduced the adhesion strength of ice. The adhesion forces of PDMS resins (Samples A, B, and D) were 291, 55, and 74 kPa, respectively [14, 15]. Our NSGs (Sample E, α = 300, composition of AR20 is 100%) and TR-SLUGs (Sample F, α = 300, compositions of TSF437 and AR20 are 25% and 75%, respectively) showed extremely low adhesion force (3.1 and 0.4 kPa, respectively) [9]. These adhesion forces were far smaller than those of the anti-icing SLIPS reported so far (Sample C, 15.6 kPa) [16]. An ice pillar that was prepared on a TR-SLUG surface could slide off from a slightly inclined surface even at −15°C. Conversely, ice pillars formed on the original PDMS resin and NSG surfaces did not slide under these conditions. In particular, an ice pillar on the PDMS resin strongly adhered, even after it was tilted to 180°. We are now investigating the antisnow properties of our NSG and TR-SLUG on solar cell panels during the winter season in Hokkaido, Japan [17].

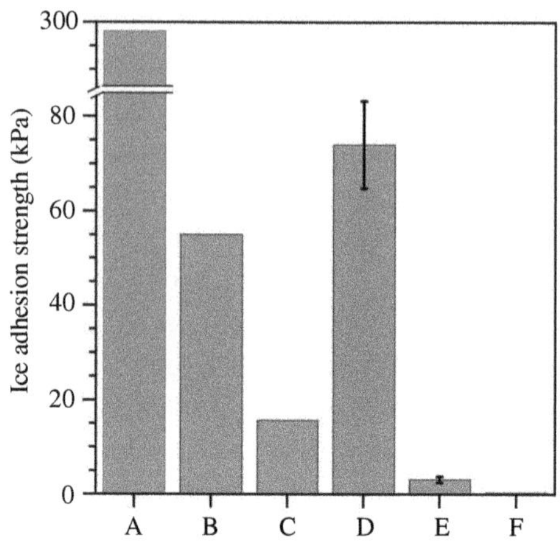

Figure 5.5 Adhesion strength between ice and various samples: (A) PDMS [14], (B) PDMS [15], (C) SLIPS [16], (D) PDMS used in this study, (E) NSG (AR20), and (F) SLUG (AR20+TSF437).

5.4 Summary

In this chapter, we introduced our novel organogels, named "self-lubricating gels" or SLUGs, showing multiple functionalities through a one-pot cross-linking reaction of PDMSs in the presence of several organic liquids. By tuning the compatibility of the PDMS matrices with the infused organic liquids, we could arbitrarily control rates of syneresis from the PDMS matrices and were also able to activate syneretic behavior by varying the surrounding conditions.

By taking advantage of this syneresis phenomenon, our SLUG surfaces can lower the adhesion strength of water, viscous liquids (even after drying), and ice/snow, through the formation of a sufficiently thick syneresis liquid layer on their topmost surfaces. In addition, through the selection of a reactive liquid, such as ODS, spontaneous superhydrophobicity can be induced on the SLUG surface due to its polycondensation reactions with ambient moisture. This regeneration of superhydrophobicity was also observed on these samples when new surfaces were created by inflicting mechanical damage, for example, by cutting with a box cutter or breaking surface microstructures with an electric drill. Other advantages of our SLUGs as opposed to conventional SLIPS and swollen gels are summarized as follows: (i) the syneresis volume of impregnated liquids can be arbitrarily tuned by simply varying the composition of the precursor mixtures, (ii) liquids/ substances with low compatibility toward the PDMS precursors can be homogeneously incorporated using organic liquids with high compatibility, (iii) there are no significant limitations for available substrates (surface modification is necessary to improve adhesion properties), (iv) neither surface roughening nor perfluorination is required, and (v) it may be possible to form large-area self-standing films. Thus, our SLUGs, with their dynamic, multiple, and self-repairing functionalities, undoubtedly show great potential for a wide variety of practical applications.

Acknowledgment

This work was partially supported by JSPS KAKENHI Grant Number Creating JP24120005 (A. Hozumi and C. Urata) in Scientific Research on Innovative Areas "Innovative Materials Engineering Based on Biological Diversity."

References

1. Wong, T.-S., Kang, S. H., Tang, S. K. Y., Smythe, E. J., Hatton, B. D., Grinthal, A. and Aizenberg, J. (2011). Bioinspired self-repairing slippery surfaces with pressure-stable omniphobicity, *Nature*, **477**, pp. 443–447.
2. Wong, T.-S., Sun, T., Feng, L. and Aizenberg, J. (2013). Interfacial materials with special wettability, *MRS Bull.*, **38**, pp. 366–371.
3. Lv, J., Song, Y., Jiang, L. and Wang, J. (2014). Bio-inspired strategies for anti-icing, *ACS Nano*, **8**, pp. 3152–3169.
4. Li, Y., Li, L. and Sun, J. (2010). Bioinspired self-healing superhydrophobic coatings, *Angew. Chem. Int. Ed.*, **49**, pp. 6129–6133.
5. Wang, X., Liu, X., Zhou, F. and Liu, W. (2011). Self-healing superamphiphobicity, *Chem. Commun.*, **47**, pp. 2324–2326.
6. Manna, U., Carter, M. C. D. and Lynn, D. M. (2013). "Shrink-to-Fit" superhydrophobicity: thermally-induced microscale wrinkling of thin

hydrophobic multilayers fabricated on flexible shrink-wrap substrates, *Adv. Mater.*, **25**, pp. 3085–3089.

7. Wu, G., An, J., Tang, X.-Z., Xiang, Y. and Yang, J. (2014). A versatile approach towards multifunctional robust microcapsules with tunable, restorable, and solvent-proof superhydrophobicity for self-healing and self-cleaning coatings, *Adv. Funct. Matter.*, **24**, pp. 6751–6761.
8. Nakajima, R. and Tanaka, Y. (2014). The role of coral mucus in the material cycle in reef ecosystems: biogeochemical and ecological perspectives, *J. Jpn. Coral Reef Soc.*, **16**, pp. 3–27 (in Japanese).
9. Urata, C., Dunderdale, G. J., England, M. W. and Hozumi, A. (2015). Self-lubricating organogels (SLUGs) with exceptional syneresis-induced anti-sticking properties against viscous emulsions and ices, *J. Mater. Chem. A*, **3**, pp. 12626–12630.
10. Okay, O. (2000). Macroporous copolymer networks, *Prog. Polym. Sci.*, **25**, pp. 711–779.
11. Daniel, D., Mankin, M. N., Belisele, R. A., Wong, T.-S. and Aizenberg, J. (2013). Lubricant-infused micro/nano-structured surfaces with tunable dynamic omniphobicity at high temperatures, *Appl. Phys. Lett.*, **102**, p. 231603 (1–4).
12. Ke, Q., Fu, W., Wang, S., Tang, T. and Zhang, J. (2010). Facile preparation of superhydrophobic biomimetic surface based on octadecyltrichlorosilane and silica nanoparticles, *ACS Appl. Mater. Interfaces*, **2**, pp. 2393–2398.
13. Ke, Q., Li, G., Liu, Y., He, T. and Li, X.-M. (2010). Formation of superhydrophobic polymerized *n*-octadecylsiloxane nanosheets, *Langmuir*, **26**, pp. 3579–3584.
14. Meuler, A. J., Smith, J. D., Varanasi, K. K., Mabry, J. M., McKinley, G. H. and Cohen, R. E. (2010). Relationships between water wettability and ice adhesion, *ACS Appl. Mater. Interfaces*, **2**, pp. 3100–3110.
15. Zhu, L., Xue, J., Wang, Y., Chen, Q., Ding, J. and Wang, Q. (2013). Ice-phobic coatings based on silicon-oil-infused polydimethylsiloxane, *ACS Appl. Mater. Interfaces*, **5**, pp. 4053–4062.
16. Kim, P., Wong, T.-S., Alvarenga, J., Kreder, M. J., Adorno-Martinez, W. E. and Aizenberg, J. (2012). Liquid-infused nanostructured surfaces with extreme anti-ice and anti-frost performance, *ACS Nano*, **6**, pp. 6569–6577.
17. http://www.nedo.go.jp/library/seika/shosai_201604/20160000000232.html

Chapter 6

Bioinspired Materials for Thermal Management Applications

Hirotaka Maeda
Department of Life Science and Applied Chemistry, Nagoya Institute of Technology, Gokiso-cho, Showa-ku, Nagoya, 466-8555, Japan
maeda.hirotaka@nitech.ac.jp

Thermal management is a key technology for saving energy in electric devices. Various types of advanced engineering materials have been prepared for thermal management applications. Bioinspired materials provide new material designs that are different from those of advanced engineering materials. This chapter introduces the use of two types of insects, the Saharan silver ant and the morpho butterfly, and diatom frustules in thermal management applications.

6.1 Introduction

Thermal management is needed to promote energy conservation to reduce the environmental load and maintain a high performance

Industrial Biomimetics
Edited by Akihiro Miyauchi and Masatsugu Shimomura

ISBN 978-981-4800-07-5 (Hardcover), 978-0-429-05883-7 (eBook)
www.jennystanford.com

for electric devices. Thermal management materials have been used in various applications [1–4]. Understanding thermal transfer is an important factor when designing materials for thermal management applications. First, heat transfer will be explained briefly. Transformation of heat, which originates from a difference in heat over a distance between two points, is categorized into three mechanisms: thermal conduction, thermal convection, and thermal radiation.

Thermal conduction is a fundamental factor in thermal transfer when solids have a heat gradient. It is predominantly carried out by phonons and free electrons in nonmetal and metal materials, respectively. In the case of liquids and gases, thermal convection acts as the dominant heat transfer mechanism based on natural and forced fluid motions. Thermal radiation is a process involving the emittance and absorption of electromagnetic waves by materials using the thermal motions of charged particles. Thermal radiation has a different nature than thermal conduction and convection. That is, thermal radiation helps in thermal transfer only under vacuum conditions.

In the past decade, advanced thermal management materials have been developed using new types of functional materials in nanometer-order sizes, such as carbon nanotubes [5–7]. Recently, bioinspired materials have received attention for use as next-generation materials in various fields [8–10]. In this chapter, studies on bioinspired thermal management materials will be introduced.

6.2 Saharan Silver Ants

Saharan silver ants (*Cataglyphis bombycina*) are found in the Sahara Desert, that is, a tropical region. Although they live under high-temperature conditions, they exhibit almost the same behaviors as ants found in other regions. This section will present the research on the thermal management strategy of Saharan silver ants [11]. The ants have silver hairs with a triangular morphology on their bodies, giving them a glossy surface, as shown in Fig. 6.1 [12]. This is caused by the local arrangement and dense packing of hairs on their backs and sides.

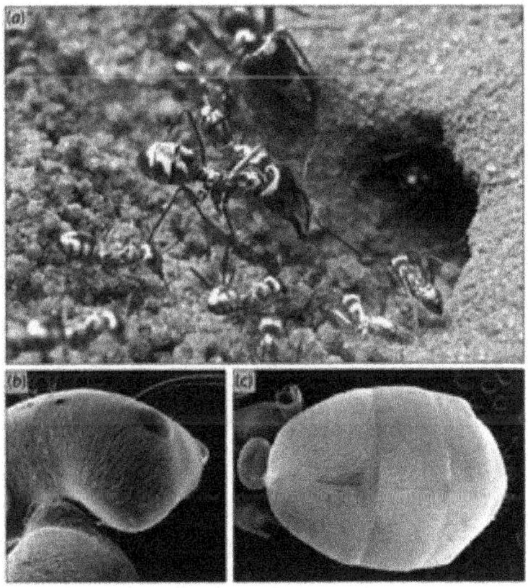

Figure 6.1 The silver ant *Cataglyphis bombycina*. Colonies contain a caste of spindly workers and a caste of soldiers with large heads and saber-shaped mandibles. (a) In full sunlight, workers and solders have a metallic sheen that justifies the vernacular name of the species. (b,c) The hairs covering the dorsal side of the workers' head, thorax (b), and abdomen (c) follow the cuticle's curvatures [12].

The reflection spectra measurements of the ants with and without their hairs indicate that the presence of the hairs leads to the development of high reflectivity in the visible-to-near-infrared range. Because of the triangular morphology of the hairs, the reflectivity increases when the incident angle of light increases from the visible range to the near-infrared range according to finite-difference time-domain simulations used for a comparative analysis between triangular and circular hair morphologies. The reason for these effects is that the hairs effectively reflect the thermal energy originating from sunlight, resulting in the suppression of thermal transfer into their body.

Interestingly, in the midinfrared range, the ants' hairs have reduced reflectivity. A decrease in the reflectivity of a material results in the enhancement of its thermal emissivity according to Kirchhoff's law for thermal emittance equilibrium. The ants exhibit vigorous activity when their body temperature is in the range of 48°C to 51°C [13]. At a surface temperature of 50°C, blackbody radiation occurs in the range of 6 to 16 μm. As a result, heat is dissipated efficiently from the body without hairs. This mechanism is an important key in designing new materials for passive radiative cooling. It is difficult to investigate the effect of the ant hairs on their thermal properties using conventional thermal equipment. Therefore, a unique device for thermal analysis was constructed, as shown in Fig. 6.2. A high-energy xenon lamp was used as a light source to mimic sunlight radiation. Thermoelectric cooling plates acted as thermal radiators under a clear sky. To minimize thermal convection under experimental conditions, the ants were placed in a vacuum chamber that contained transmission windows in the visible range and the infrared range to measure temperature changes using thermography. Human hair was a unique feature used to minimize thermal conduction. This experimental system can be used to evaluate the effects of thermal radiation and those of natural convention and thermal radiation under vacuum and ambient atmospheres, respectively.

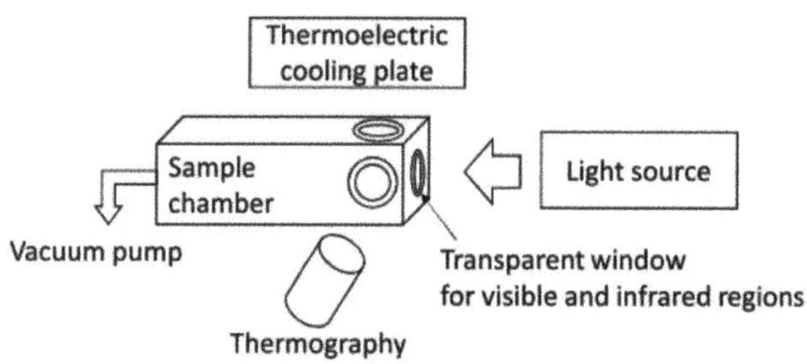

Figure 6.2 Schematic illustration of the thermal analysis equipment for silver ants.

Another approach to evaluating ant body temperature is to use a micrometer-sized thermocouple inserted into the ant's rectum. Thermodynamics experiments clearly showed that the ants' hairs suppressed increases in body temperature (Fig. 6.3). The surface properties of the hairs play an important role in tuning their thermal management.

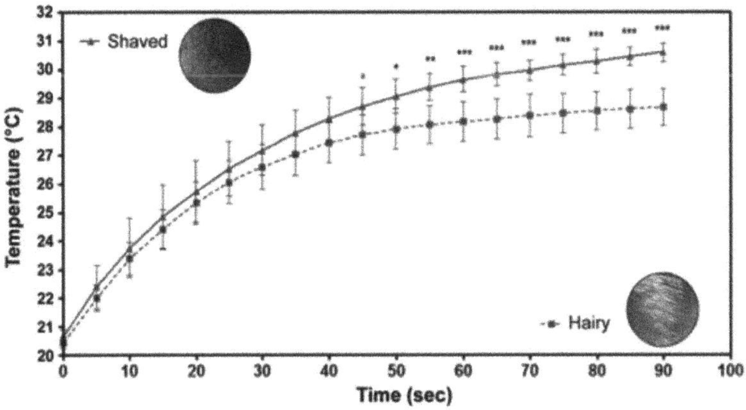

Figure 6.3 Heating curves of hairy and shaved *C. bombycina* workers. Abdomens were exposed for 90 s to light emitted by a solar simulator whose visible spectral intensity was identical to that of the summer Saharan sun. The difference in excess temperature became significant after 45 s; after 90 s of exposure, the difference in excess temperature between both conditions reached 1.91°C. Unpaired Student's *t*-test ($^*p < 0.05$, $^{**}p < 0.01$, $^{***}p < 0.001$) [12].

6.3 Morpho Butterfly

In thermal management applications, thermal detection and imaging are also important subjects in resolving energy and environment problems. In general, the thermal energy of radiation emitted from materials can be visualized as temperature using infrared thermography utilizing microbolometers, which are widely used in various fields. Recently, new engineering approaches for thermal detection and imaging using thermal bimorphs and Fabry–Perot interferometers have been proposed [14, 15].

The wings of the morpho butterfly exhibit a multilayer interference of light due to their nanoarchitecture, leading to the expression of iridescent colors [16]. Uses of the natural photonic morpho nanostructure for thermal management applications will be described in this section. Potyrailo et al. proposed a new concept for an infrared detector inspired by the iridescent morpho butterfly (Fig. 6.4) [17]. To develop the new type of thermal detector, the morpho butterfly wing was doped with single-walled carbon nanotubes. This led to the upconversion of midwave infrared radiation in the range of 3–8 μm to iridescent visible waves through the modulation of the optical cavity due to changes in thermal expansion and the refractive index.

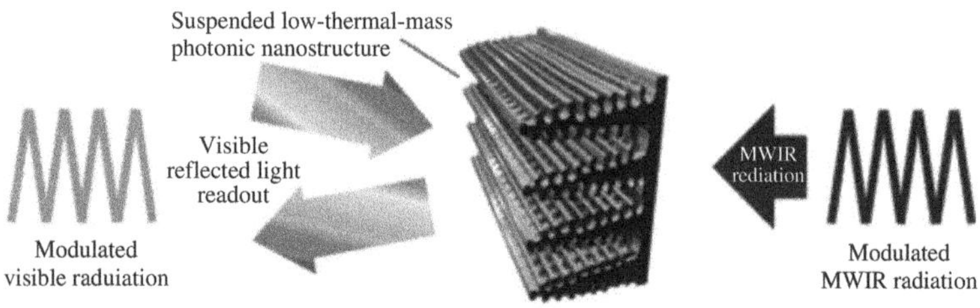

Figure 6.4 New principle for uncooled thermal detection based on the air-filled photonic architecture of the iridescent scales of the Morpho butterfly. Experimental schematic with a free-standing Morpho nanostructure, its excitation with midwave infrared (MWIR) radiation, and the readout of the visible response. Reproduced from Ref. [17] with permission. Copyright 2012, Nature Publishing Group.

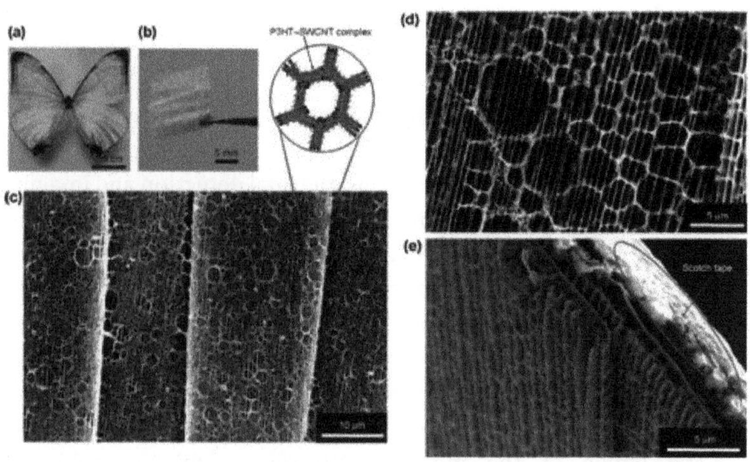

Figure 6.5 Structural characterization of the carbon nanotube (CNT)-butterfly wing composite. (a) Image of the *M. sulkowskyi* butterfly. (b) Image of the CNT-butterfly wing composite. (c, d) SEM images of the CNT-butterfly wing composite and a close-up view. Inset: Schematic illustration of the honeycomb-shaped microstructure based on poly(3-hexylthiophene) (P3HT)-SWCNT complexes. (e) SEM images of the cross section of the CNT-butterfly wing composite. Scotch tape was used to reinforce the composite during the observation of the structures by SEM. Reprinted with permission from Ref. [18]. Copyright (1969) American Chemical Society.

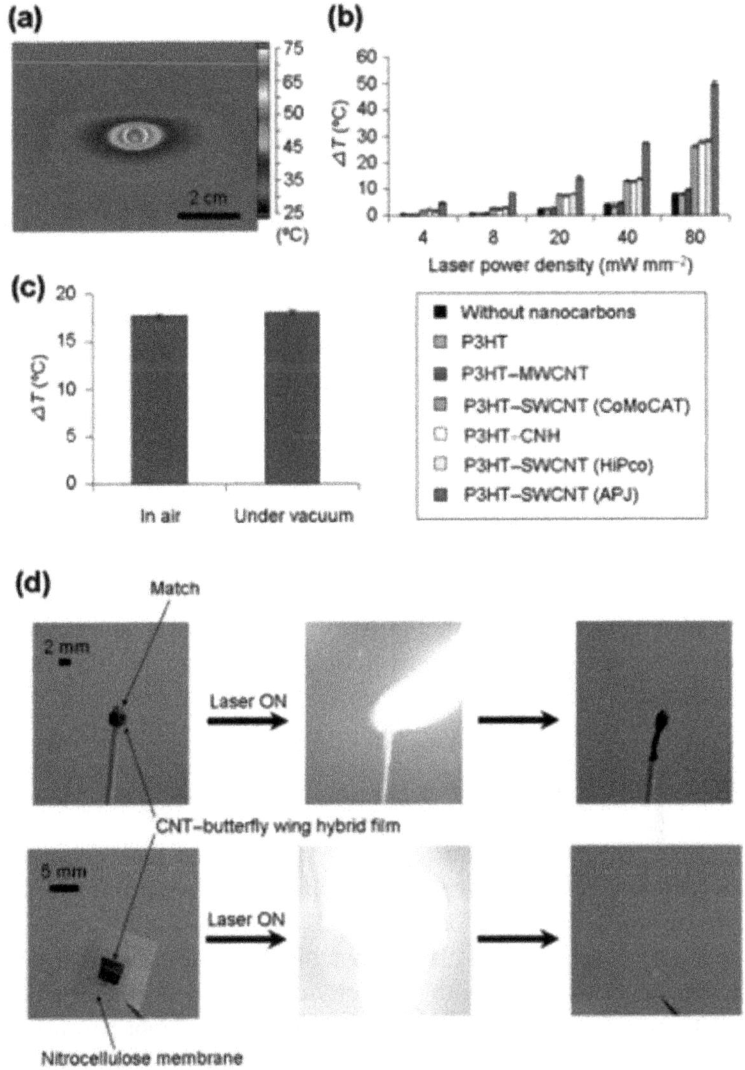

Figure 6.6 Photothermal conversion of the CNT-butterfly wing composite. (a) Thermographic images of the surface of the CNT-butterfly wing composite using a 785 nm NIR laser at 1000 mW (ca. 80 mW/mm^2). (b) Photoinduced temperature elevation on the composites at various laser powers (ca. 4, 8, 20, 40, and 80 mW/mm^2). (c) Photoinduced temperature changes on the composites using a 785 nm NIR laser at 1000 mW (ca. 80 mW/mm^2) in air or under vacuum. (d) Fire tests of the photoinduced CNT-butterfly wing composite. Laser wavelength = 785 nm. Laser power = 1000 mW (ca. 80 mW/mm^2). Reprinted with permission from Ref. [18]. Copyright (1969) American Chemical Society.

Another exciting study on morpho butterfly wings combined with single- or multiwalled carbon nanotubes for thermal management applications was reported by Miyako et al. [18]. This type of composite was prepared by forming a honeycomb-shaped network containing various types of carbon nanotubes on the wing using a self-assembly technique, as shown in Fig. 6.5.

The photothermal effects of the composites were examined using laser irradiation at 785 nm, because the honeycomb-like pattern of the same butterfly wing scales takes advantage of refraction to trap light, much like a fiber-optic cable [19]. The wings combined with carbon nanotubes enhanced heat induction by the laser, dependent on the carbon nanotube type (Fig. 6.6a,b). This effect had no influence on the atmosphere, which was related to thermal convection (Fig. 6.6c). The laser irradiation of the composite with a honeycomb structure induced excessive heat, resulting in the ignition of a match or nitrocellulose membrane (Fig. 6.6d). On the other hand, in the case of the wings or carbon nanotubes alone, laser irradiation did not result in the ignition of a match. The unique nanostructures of morpho butterfly wings combined with engineered nanomaterials have great potential for thermal management applications and developing novel functionalizations.

6.4 Diatom Frustule

Diatoms are photosynthetic micro-organisms, and their frustules have unique species-dependent mesoporous structures [20]. The hydrothermal treatment of compacts consisting of diatomaceous earth composed chiefly of diatom frustules has been used to prepare mesoporous materials with humidity-controlling properties [21, 22]. Diatom frustules can adsorb and desorb water vapor because of their porous properties. The effect of water vapor adsorbed into the mesopores of diatom frustules on their thermal properties [23] will be explained in this section.

Heat capacity ($J \cdot K^{-1}$), which is the quantity of thermal energy required to increase the temperature of a material by 1 K, is a key thermal property of materials. It is well known that water has a high specific heat among liquids, which is related to heat capacity. The use of water is important in thermal management applications, such as water-cooling systems. When water is used for thermal management, it must be retained on the inside of the materials. The introduction of adequate mesopores into a material could cause water vapor to condense from the atmosphere at an appropriate relative humidity without additional energy. This is explained by the Kelvin equation (Eq. 1.1) for capillary condensation [24],

$$r = \frac{2\gamma M \cos\theta}{\rho RT \ln(P/P_0)}, \tag{6.1}$$

where r is the pore size, γ is the surface tension of water, M is the molecular weight of water, θ is the contact angle, ρ is the density of water, R is the gas constant, T is the absolute temperature, and P/P_0 is the relative water vapor pressure. The relationship between pore size and relative humidity is shown in Fig. 6.7. Mesoporous materials derived from silica, clays, and minerals can regulate the humidity due to capillary condensation [25–27]. In Japan, kura, traditional storehouses, can keep foods fresh by controlling the indoor humidity through capillary condensation by utilizing the mesopores of the soil's aggregation structure.

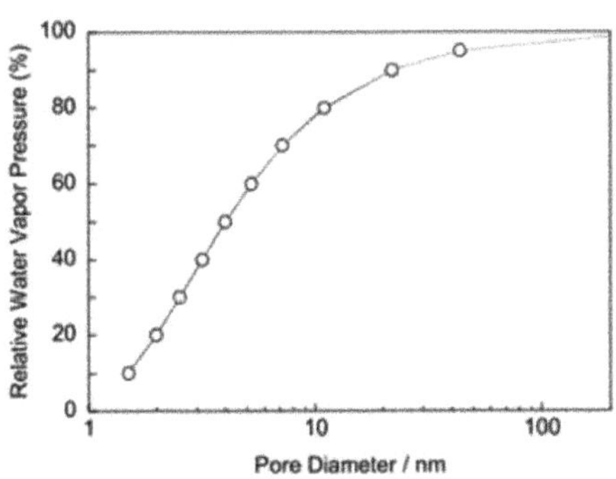

Figure 6.7 Relationship between relative water vapor pressure and pore diameter in the case of a zero of contact angle.

Diatom frustules were obtained from *Fistulifera solaris* using thermal and acid treatments (Fig. 6.8). The treatments cause morphological changes in the frustules, resulting in changes in the porous properties; the specific surface areas after thermal and acid treatments were 100 and 230 m^2/g, respectively. The acid-treated frustules had a wide pore size distribution (in the range of 3 to

20 nm). On the other hand, the thermal-treated frustules had a sharp peak at 4 nm in the pore size distribution curve. The mesopore volume of the acid-treated frustules was higher than that of the thermal-treated frustules. The water vapor adsorption-desorption isotherms of these frustules indicate that the acid-treated samples have a better water vapor adsorption ability than the thermal-treated samples (Fig. 6.9), which is consistent with the prediction using the Kelvin equation for capillary condensation.

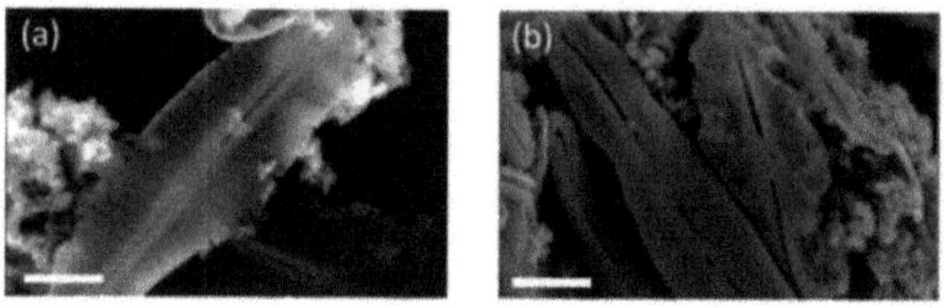

Figure 6.8 SEM images of diatom frustules after (a) acid and (b) heat treatments. Scale bar = 1 μm. From Ref. [23]. Copyright 2017. With permission of Springer.

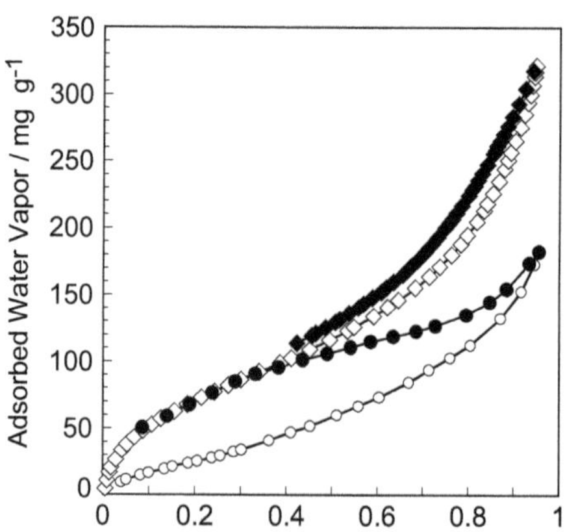

Figure 6.9 Water vapor isotherms for (square) acid- and (circle) thermal-treated diatom frustules at 25° C. Open and filled marks represent adsorption and desorption branches, respectively. From Ref. [23]. Copyright 2017. With permission of Springer.

The thermal analysis results show an increase in the temperature increase rate with the adsorption of water vapor, independent of the treatment method. The temperature increase rate tends to increase with the amount of adsorbed water vapor. The increase in the amount of adsorbed water vapor decreases in the volume of air in the mesopores of the frustules (Fig. 6.10). It is well known that water has a higher thermal conductivity than air. The condensed water vapor due to adsorption in the mesopores behaves as a thermal transfer pathway, resulting in improved thermal properties. The combination of diatom frustules and the adsorption of water vapor into the mesopores has the potential for novel thermal management applications.

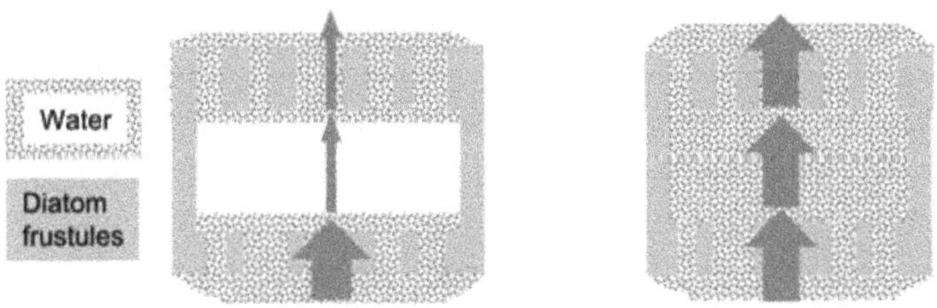

Figure 6.10 Theoretical drawings of diatom frustules with different amounts of adsorbed water vapor. Arrows represent thermal conduction. Residual air in the frustules reduces thermal transfer.

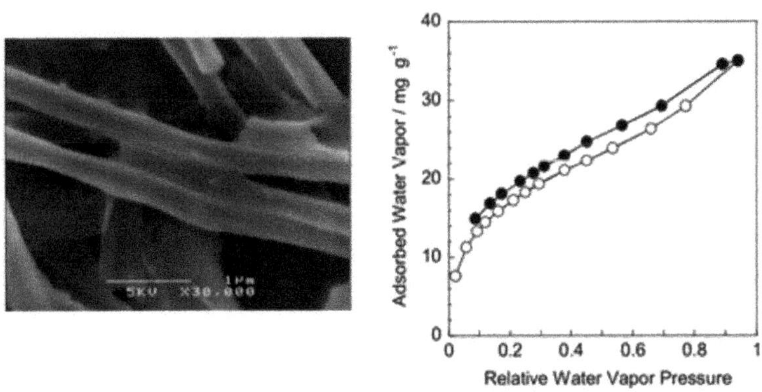

Figure 6.11 (a) SEM image and (b) water vapor isotherms at 25°C for acid-treated diatom frustules derived from *Chaetoceros gracilis*.

Chaetoceros gracilis was used to investigate the effect of morphology on thermal properties. *C. gracilis* frustules have a morphology different from that of *F. solaris* (Fig. 6.11), and the mesoporous properties of the former, such as specific surface area and mesopore volume, are inferior. *C. gracilis* frustules have a much lower water vapor adsorption ability than *F. solaris* frustules, resulting in inferior thermal transfer properties. The water vapor adsorption abilities, which depend on the mesoporous properties of the frustules, contribute strongly to the thermal transfer properties.

6.5 Summary

In this chapter, various bioinspired materials for thermal management applications were introduced. Saharan silver ants can adjust to thermal radiation due to the morphology and arrangement of their hairs, by suppressing the increase in their body temperature. The combination of the unique nanostructures and natural photonic structures of the wings of the morpho butterfly and carbon nanotubes has possible novel thermal management applications, such as in thermal detection and diagnosis. The adsorption of water vapor from the ambient atmosphere without the need for additional energy into the mesopores of diatom frustules enhances the thermal transfer properties. Thus, bioinspired materials are novel candidate materials for thermal management applications.

Acknowledgments

This work was supported partially by JSPS KAKENHI Grant Number JP24120005 in Scientific Research on Innovative Areas "Innovative Materials Engineering Based on Biological Diversity."

References

1. Matsunaga, T., Yamada, N., Kojima, R., Shamoto, S., Sato, M., Tanida, H., Uruga, T., Kohara, S., Takata, M., Zalden, P., Bruns, G., Sergueev, I., Wille H. C., Hermann, R. P. and Wuttig, M. (2001). Phase-change materials: vibrational sogtening upon crystallization and its impact on thermal properties, *Adv. Funct. Mater.*, **21**, pp. 2232–2239.
2. Wang, Z., Carter, J. A., Lagutchev, A. and Dlott, D. D. (2007). Ultrafast flash thermal conductance of molecular chains, *Science*, **317**, pp. 787–790.

3. Chang, C. W., Okawa, D., Majumdar, A. and Zettl, A. (2006). Solid-state thermal rectifier, *Science*, **314**, pp. 1121–1124.
4. Cahill, D. G., Ford, W. K., Goodson, K. E., Mahan, G. D., Majumdar, A., Maris, H. J., Merlin, R. and Phillpot, S. R. (2003). Nanoscale thermal transport, *J. Appl. Phy.*, **93**, pp. 793–818.
5. Stankovich, S., Dikin, D. A., Dommett, G. H. B., Kohlhaas, K. M., Zimney, E. J., Stach, E. A., Piner, R. D., Nguyen, S. T. and Ruodd, R. S. (2006). Graphene-based composite materials, *Nature*, **442**, pp. 282–286.
6. Han, Z. D. and Fina, A. (2011). Thermal conductivity of carbon nanotubes and their polymer nanocomposites: a review, *Prog. Poly. Sci.*, **36**, pp. 914–944.
7. Balandin, A. A., Ghosh, S., Bao, W. Z., Calizo, I., Teweldebrhan, D., Miao, F. and Lau, C. N. (2008). Superior thermal conductivity of single-layer graphene, *Nano Lett.*, **8**, pp. 902–907.
8. Wong, T. S., Kang, S. H., Tang, S. K. Y., Smythe, E. J., Hatton, B. D., Grinthal, A. and Aizenberg, J. (2011). Bioinspired self-repairing slippery surfaces with pressure-stable omniphobicity, *Nature*, **477**, pp. 443–447.
9. Darder, M., Aranda, P. and Ruiz-Hlitzky, E. (2007). Bionanocomposites: a new concept of ecological, inspired, and functional hybrid materials, *Adv. Mater.*, **19**, pp. 1309–1319.
10. Bonderer, L. J., Studart, A. R. and Gauckler L. J. (2008). Bioinspired design and assembly of platelet reinforced polymer films, *Science*, **319**, pp. 1069–1073.
11. Shi, N. N., Tsai, C.-C., Camino, F., Bernard, G. D., Yu, N. and Wehner, R. (2015), Keeping cool: enhanced optical reflection and radiative heat dissipation in Saharan silver ants, *Science*, **349**, pp. 298–301.
12. Willot, Q., Simonis, P., Vigneron, J.-P. and Aron, S. (2016). Total internal reflection accounts for the bright color of the Saharan silver ant, *PLoS One*, **11**, p. e0152325.
13. Wehner, R., Marsh, A. C. and Wehner, S. (1992). Desert ants on a thermal tightrope, *Nature*, **357**, pp. 586–587.
14. LeMieux, M. C., McConney, M. E., Lin, Y.-H., Singamaneni, S., Jiang, H., Bunning, T. J. and Tsukruk, V. V. (2006). Polymer nanolayers as actuators for ultrasensitive thermal bimorphs, *Nano Lett.*, **6**, pp. 730–734.
15. Choi, H. Y., Park, K. S., Park, S. J., Paek, U. C., Lee, B. H. and Choi, E. S. (2008). Miniature fiber-optic high temperature sensor based on a hybrid structured Fabry-Perot interferometer, *Opt., Lett.*, **33**, pp. 2455–2457.

16. Vukusic, P., Sambles, J. R., Lawrence, C. R. and Wootton, R. J. (1999). Quantified interference and diffraction in single Morpho butterfly scales, *Proc. R. Soc. London, Ser. B*, **266**, pp. 1403–1411.

17. Pris, A. D., Utturkar, Y., Surman, C., Morris, W. G., Vert, A., Zalyubovskiy, S., Deng, T., Ghiradella, H. T. and Potyrailo, R. A. (2012). Towards high-speed imaging of infrared photons with bio-inspired nanoarchitectures, *Nat. Photonics*, **6**, pp. 195–200.

18. Miyako, E., Sugino, T., Okazaki, T., Bianco, A., Yudasaka, M. and Iijima S. (2013). Self-assembled carbon nanotube honeycomb networks using a butterfly wing template as a multifunctional nanobiohybrid, *ACS Nano*, **7**, pp. 8736–8742.

19. Zhang, W., Zhang, D., Fan, T., Gu, J., Ding, J., Wang, H., Guo, Q. and Ogawa, H. (2009). Novel photoanode structure templated from butterfly wing scales, *Chem. Mater.*, **21**, pp. 33–40.

20. Round, F. E., Crawford, R. M. and Mann D. G. (2007). *Diatoms: Biology and Morphology of the Genera* (Cambridge University Press, Cambridge).

21. Jing, Z., Maeda, H., Ioku, K. and Ishida, E. H. (2007). Hydrothermal synthesis of mesoporous materials from diatomaceous earth, *AIChE J.*, **53**, pp. 2114–2122.

22. Maeda, H., Kato, S. and Ishida, E. H. (2009). Preparation of hydrothermally solidified mesoporous materials from diatomaceous earth for moisture control application, *Int. J. Appl. Ceram. Technol.*, **6**, pp. 431–436.

23. Maeda, H., Matsumoto, M., Maeda, Y., Egashira, Y., Tanaka, T. and Kasuga, T. (2017). Utilization of diatom frustules for thermal management applications, *J. Appl. Phycol.*, **29**, pp. 1907–1911.

24. Fisher, L. R., Gamble, R. A. and Middlehurst, J. (1981). The Kelvin equation and the capillary condensation of water, *Nature*, **290**, pp. 575–576.

25. Cagnol, F., Grosso, D., Solerr-Illia, G. J. D. A. S., Crepaldi, E. L., Babonneau, F., Amenitsch, H. and Sanchez, C. (2003). Humidity-controlled mesostructuration in CTAB-templated silica thin film processing. The existence of modulable steady state, *J. Mater. Chem.*, **13**, pp. 61–66.

26. Hemmen, H., Alme, L. R., Fossum, J. O. and Meheust, Y. (2010). X-ray studies of interlayer water absorption and mesoporous water transport in a weakly hydrated clay, *Phy. Rev. E*, **82**, p. 036316.

27. Maeda, H. and Ishida, E. H. (2009). Water vapor adsorption and desorption of mesoporous materials derived from metakaolinite by hydrothermal treatment, *Ceram. Int.*, **35**, pp. 987–990.

Chapter 7

Strange Wing Folding in a Rove Beetle

Kazuya Saito
Faculty of Design, Kyushu University, Shiobaru 4-9-1, Minami-ku, Fukuoka 815-0032, Japan
k-saito@design.kyushu-u.ac.jp

7.1 Introduction

Technologies to fold large-sized structures into compact shapes are required in a variety of engineering applications in multiple scales and for multiple purposes, such as space-deployable structures represented by solar array paddles and antenna reflectors of satellites, medical devices, and many articles of daily use, like fans and umbrellas (Fig. 7.1). Insects have addressed the same issue for hundreds of millions of years as they acquired flight wings and developed several interesting solutions. The acquisition of storing function in the wing was an important evolutionary event, one comparable to the acquisition of the wing itself. Foldable wings ironed out the disadvantages of the wing, bulkiness and vulnerability, and gave the abilities of high-speed locomotion on ground and penetration into narrow spaces, like the leaf litter layer

Industrial Biomimetics
Edited by Akihiro Miyauchi and Masatsugu Shimomura

ISBN 978-981-4800-07-5 (Hardcover), 978-0-429-05883-7 (eBook)
www.jennystanford.com

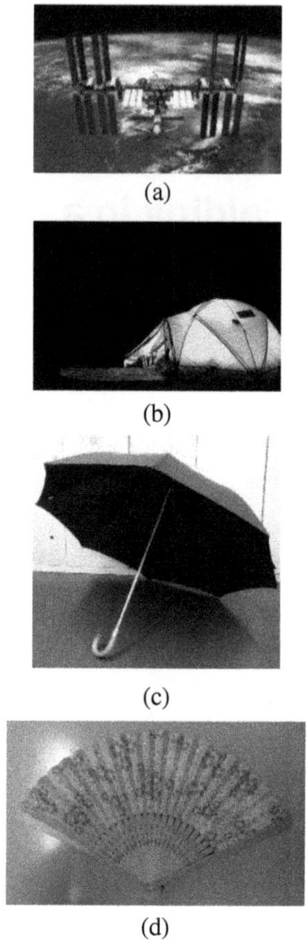

Figure 7.1 Deployable structures in various scales and for various purposes. (a–d) Space-deployable structures. (a) Solar array paddles in the International Space Station. (b) A tent. (c) An umbrella. (d) A folding fan.

and the area under the soil, without loss of flight ability, enabling fast and long-range migration. In the same way, folding technologies are expected to generate innovative ideas for usage and application in various engineering products. From an engineering standpoint, these foldable wings have an interesting property: they can enable compact storing without compromising the strength and stiffness of the wing structure. It is well known that many species of insects are equipped with deployable wings. Simple examples are found in the fan-like folding of locusts (Orthoptera) and praying mantises

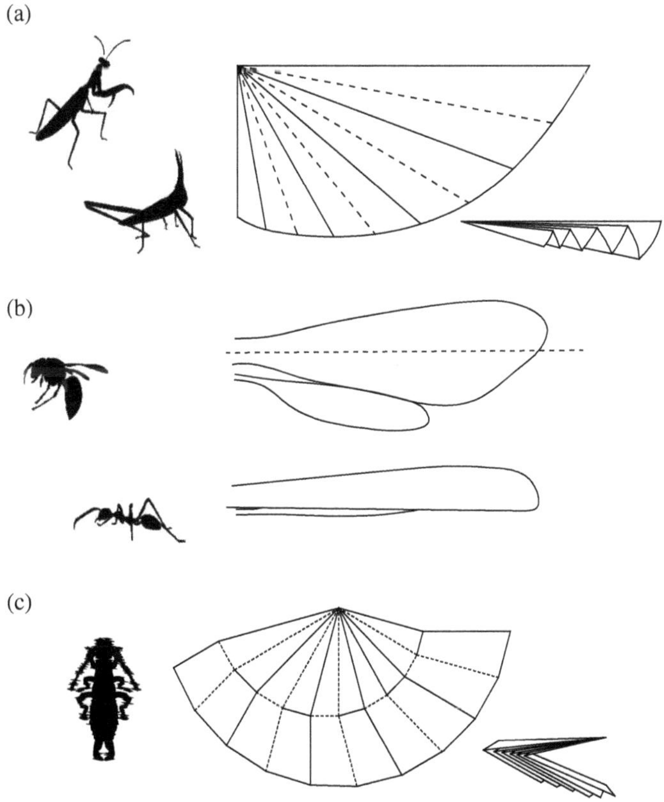

Figure 7.2 Representative folding methods in insect wings (not involving beetles). (a) Fan-like folding in locusts (Orthoptera) and praying mantises (Dictyoptera) [2]. (b) Simple longitudinal folding line in ants and bees (Hymenoptera) [1]. (c) Advanced fan-like folding in earwigs (Dermaptera) [5].

(Dictyoptera) [1] (Fig. 7.2a). If you carefully observe a bee landing on a flower, you will find that its forewings are folded along the width direction [2] (Fig. 7.2b). These types of deployable wings have simple crease patterns and can be folded and unfolded by relatively easy mechanisms; however, their storage efficiencies are not large. Earwigs (Dermaptera) are known to use a unique wing-folding method. Their wings have fan-like folding patterns, but the fan frames have additional bending points, and earwigs achieve more extensive folding by refolding the closed fan [3, 4] (Fig. 7.2c). All these examples are very interesting research targets from the point of view of biomimetics, but the most highly diverse wing-folding patterns and mechanisms are found in beetles (Coleoptera). Depending on the shape of the abdomen and the survival strategy, each species has selected an optimal crease pattern and folding/unfolding mechanism. Among these examples, the most sophisticated wing folding is found in rove beetles (Coleoptera: Staphylinidae). In addition to the best performance as a deployable structure, the wings of rove beetles have a strange feature that is not observed in other beetles; their right and left wings use different folding patterns [5]. This chapter focuses on this strange wing folding in rove beetles and shows its interesting properties from the point of view of engineering applications.

7.2 Wing Folding in Beetles

7.2.1 Crease Patterns

The majority of the beetle hindwing comprises a relatively soft membrane. During flights, the wing shape is maintained by thick frames called veins. The pattern of the wing frame, venation, is an important characteristic for taxonomy and phylogeny and therefore has been investigated in detail. On the other hand, the way of wing folding has tended to be neglected by entomologists, and few studies have been devoted to this subject. This field was explored by the pioneering work of Forbes in the 1920s [6, 7]. He investigated 125 wing-folding patterns of beetles and illustrated their crease patterns. In addition to anecdotal studies, he proposed a generalized crease pattern based on his enormous observations. Figure 7.3 is the paper model of his hypothetical wing. In this model, angles of crease lines are modified for easy folding, so the model can be made by folding actual paper.

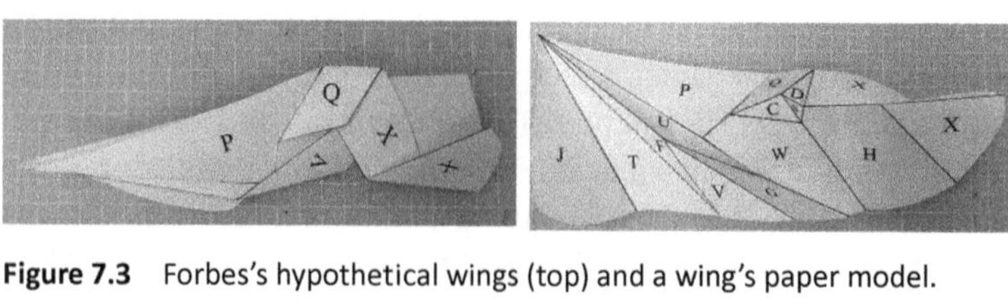

Figure 7.3 Forbes's hypothetical wings (top) and a wing's paper model.

7.2.2 Mechanisms

Entomologists have also studied the mechanisms used in wing-folding/wing-unfolding, as well as the crease patterns of the wings, and made many interesting discoveries. Unlike bird wings, insect wings do not have their roots in limbs but in the apophysis on the dorsal exoskeleton, so they cannot be actuated only by muscular driving. Their actuations are the results of the combination of various mechanisms. Representative factors are as follows.

7.2.2.1 Wing elasticity

Dissecting a freshly killed specimen of a small beetle like a ladybird beetle, you may notice that the hindwings have elasticity in terms of the unfolding direction. If you are not careful, the hindwings are automatically deployed with easy stimulation, and it will be very difficult to fold them again. Many entomologists have pointed out this intrinsic elasticity of the hindwing plays an important role in the wing-folding/wing-unfolding mechanism [8, 9]. The hindwings in many small beetles have unfolding elasticity and are deployed automatically. This type of wing can be deployed quickly using elastic force. For example, Fig. 7.4 shows slow-motion movies of the take-off motions in a ladybird beetle and a rove beetle, which were recorded at 3000 frames per second (fps). They take only 0.1 s to deploy wings. On the other hand, some beetles have automatically folded wings because of wing elasticity. Some entomologists focused on these elasticities and proposed classification of wing-folding mechanisms.

7.2.2.2 Muscle

Off course, muscles are an indispensable factor in wing-folding/wing-unfolding actuations, but this does not mean that they control all wing frame movements. In foldable hindwings, the veins have many articuli and hinges, but no muscle is observed on these inner bending points. Wing actuating muscles are concentrated on wing bases. They mainly work for flapping in flights, but these basal muscular actions have been shown to be used in wing folding/unfolding [10].

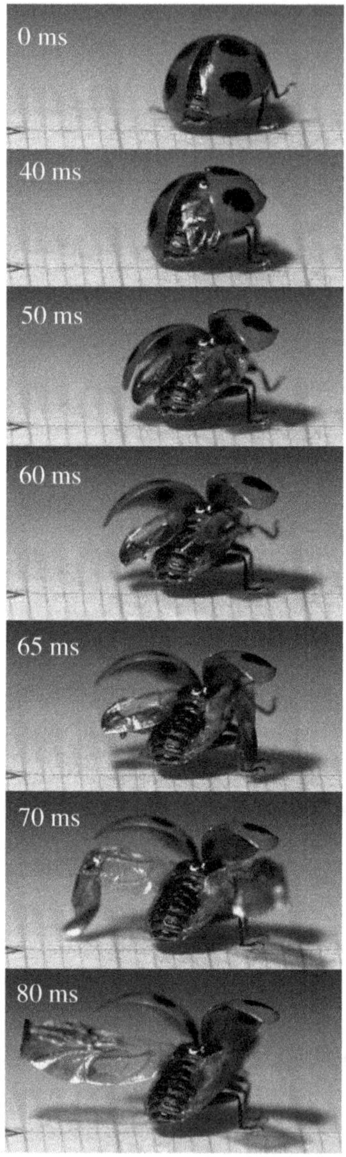

Figure 7.4 Elastic wing deployment in a ladybird beetle (*Coccinella septempunctata* L.). High-speed cameras (Phantom V1611, Novi Tech Co., Ltd., Tokyo) were used to acquire slow-motion movies (3000 fps, shutter speed 50 μs).

7.2.2.3 Other body parts

Many beetles use other body parts for wing folding. In the folding of unfolding-stable-type wings, insects must fold against the elastic force acting in the deploying direction. This elastic force is so strong that muscle actuation alone cannot accomplish the folding, hence support is needed from body parts represented by the abdomen and the elytra.

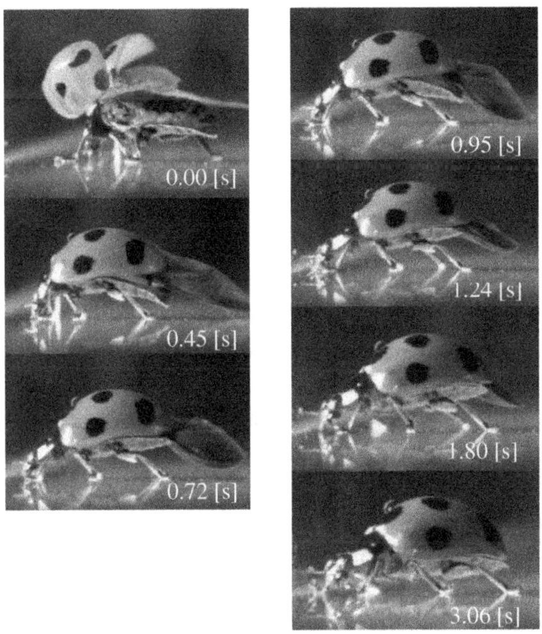

Figure 7.5 Wing-folding movement in a ladybird beetle. A high-speed camera (HAS-L2, DITECT Co., Ltd., Tokyo, Japan) was used to acquire slow-motion movies at 60 fps.

For instance, ladybird beetles fold their wings by "pumping" movements [9] of the abdomen (Fig. 7.5). An extreme example can be found in rove beetles, discussed in this chapter, which have freely movable abdomens and fold their wings in a more direct way. These types of beetles have microspicules on their tergal plates. These spicules are believed to support wing folding by acting as barbs and helping the abdomen to catch the hindwings in folding movements. The elytra, which are one of the most vital aspects of the beetle anatomy, also support wing folding by holding the midfolded hindwings in many species. If the elytra are lost, these beetles cannot fold their hindwings. The elytra in some beetles also seem to function as a hold-on release mechanism for elastically deploying hindwings. Among other potential factors is the blood pressure, which may straighten the hindwings [11]. A similar hydraulic mechanism is known to be used for wing inflation during emergencies [12].

7.3 Wing Folding in a Rove Beetle

7.3.1 What Is a Rove Beetle?

Maybe most of the readers have not heard the name "rove beetle," but they are very common insects that can be seen in various area, including in the woods, along the coast, and in the urban zone. Figure 7.6 shows some species of rove beetles. Their silhouettes look like ants in shape, but rove beetles belong to Coleoptera and are in the same group as ladybird beetles and stag beetles. Compared with other typical beetles, the elytra of most rove beetles are reduced, and the projecting abdomen is exposed and freely movable. At the expense of protection of the abdomen, rove beetles have highly maneuverable bodies that can move rapidly through narrow and curved spaces, extending their habitats to include a wide range of microhabitats, especially the leaf litter layer and the soil. The interesting thing is that they have never lost their flight wings, despite a reduction in the dorsal storage space (Fig. 7.7). Because of their short elytra, rove beetles have to fold their hindwings into more compact shapes than other typical beetles. This difficult problem is solved by their extraordinary right-left asymmetric wing folding. The folding pattern is very complex, but rove beetles have a peculiar tool that helps them fold their wings in this pattern—a movable abdomen. As a result, rove beetles became a highly diverse group, accounting for 15%, (i.e., nearly 60,000 species) of all known species of Coleoptera. The next section explains the wing-folding/wing-unfolding mechanisms in beetles explored in entomological works and challenges the clarification of this strange wing folding by using high-speed camera.

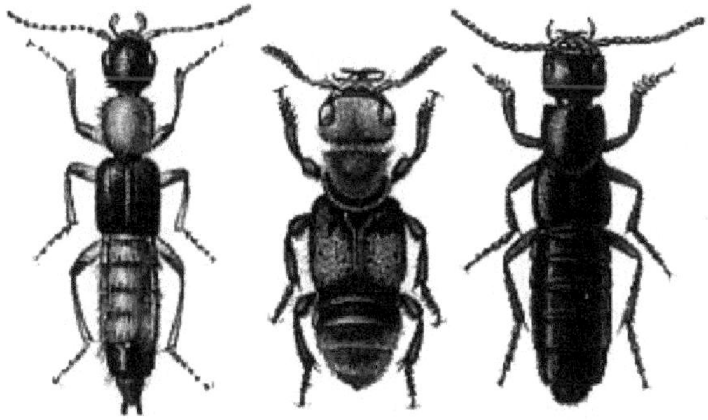

Figure 7.6 Examples of rove beetle species. (From left to right) *Paederus riparius* (7.5–8 mm), *Emus hirtus* (18–28 mm), and *Ocypus olens* (22–32 mm).

Figure 7.7 *Cafius vestitus* (Sharp). This rove beetle can be found under sea wracks on a sand beach. The flight wing can be stored beneath the elytra in an asymmetric crease pattern.

7.3.2 Wing-Folding Process in Rove Beetles

The top-left picture in Fig. 7.8 shows both wings from the body of a rove beetle dissected in the shape they are stored in. As mentioned above, the right and left wings are obviously folded in different patterns. As the wings are unfolding little by little under the microscope, the positions of the crease lines are investigated and the crease patterns of Fig. 7.8 are obtained. It can be said that wings folding in an asymmetric right-left pattern is an unexpected idea for specialists on aeronautics. When taking aerodynamic properties into account, it ideally should be better to make wings as bilaterally symmetrical as possible. Then why do rove beetles fold their wings in such a strange way? To find the answer, one has to see the actual folding method used by living rove beetles.

To investigate the detailed folding process, a camera with high-speed shooting functions is needed. Compared with wing unfolding, the folding motion is slower; so a consumer-grade camera with a 60–120 fps shooting speed is enough to observe the mechanism. Because camera technology is advancing rapidly these days, some smartphones also can be used in this experiment.

The difficulty of this experiment is summarized in the question, how can one ensure that the rove beetle flies and lands in front of the lens? To increase the success rate, it is important to select a species with high flying motivation. A coast-living rove beetle like *Cafius vestitus* (Sharp) is a good choice. Because the beetle walks around briskly, it is very difficult to chase it through a camera lens. One good method is to put the beetle in a transparent acrylic box. By moving the box in the opposite direction to the one the beetle walks in, you can always keep the specimen in focus. Marking the center of the shooting range on the desk is also very useful to keep capturing the insects by a camera.

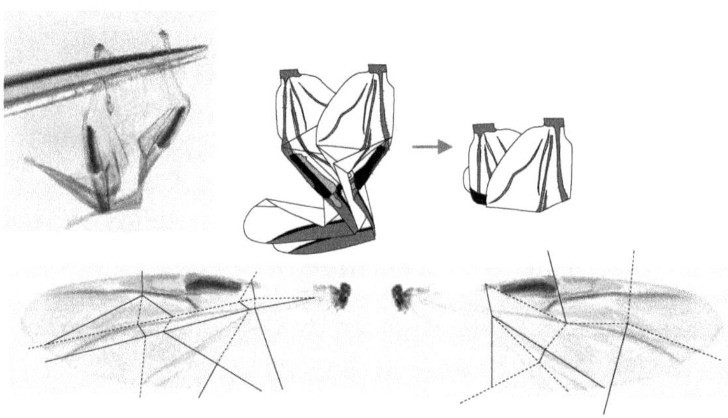

Figure 7.8 Crease patterns in *C. vestitus* (Sharp). The left and right wings are folded with using different crease patterns. In the housed state, the apexes of right and left wings are aligned to either side and folded into a Z shape (top-right figure).

Figure 7.9 shows the wing-folding motion shot at 500 fps. This motion explains why the wing folding is asymmetric. In the wing-folding processes of other insects, the left and right wings overlap after the folding process. However, the left and right hindwings of rove beetles overlap and fold simultaneously. Because of the mirror symmetry of right and left wings, the resulting crease patterns of each wing should be asymmetric. The area of the stored wing becomes 10% of the deployed wing, which is very compact storing compared with that in other beetles.

Seeing successive pictures, it is clear that the folding motion consists of multiple steps. Before they are folded, the hindwings align backward along the abdomen. Overlapped wings are folded into a perpendicular direction by the swinging movement of the abdomen from the side of the overriding wing to that of the underriding wing. Wings are then tucked beneath the elytra by the lifting movement of the abdomen and folded into a Z shape simultaneously. In the housed state, the apexes of the right and left wings are aligned to either side. This folding process is completed within about 1 s.

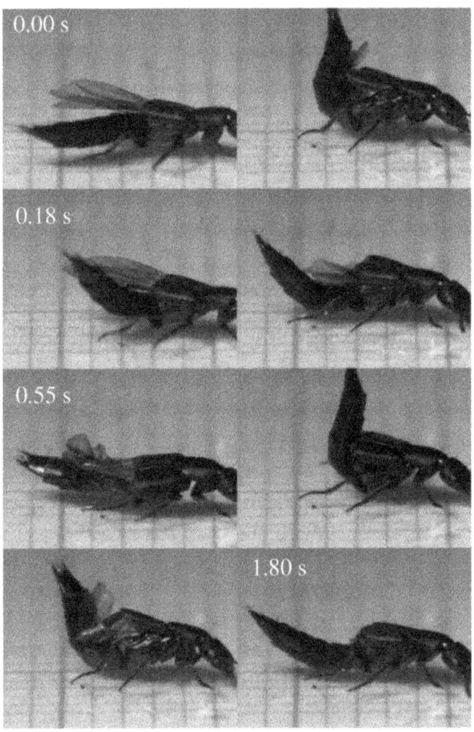

Figure 7.9 Wing-folding motion in *C. vestitus* (Sharp) imaged with a high-speed camera (HAS-L2, DITECT Co., Ltd., Tokyo, Japan).

7.3.3 From Right or Left?

To find right-left asymmetry in animal shapes is not uncommon, represented by claws in a fiddler crab, the body of a flatfish, the rolling direction of a snail's shell, etc. However, asymmetric wing folding in a rove beetle has an additional interesting feature. In Figs. 7.8 and 7.9, the rove beetle overlaps its right wing on its left wing and folds them from the right side. Is this folding direction fixed at all times? If you catch many rove beetles and see their folded wings, you will find individuals folding wings in opposite directions. In these opposite-folding cases, the left wing overlaps the right wing first and then the wings are folded from the left side by the swing movement of the abdomen. The crease pattern and the stored shape show a mirror-inversed image of Figs. 7.8 and 7.9. This fact implies either that there are individual differences in the folding direction, just like there are right-handed and left-handed humans, or that all individual rove beetles can fold their wings from both right and left sides. From an engineering viewpoint, the former is more conceivable because the number and positions of crease lines in right- and left-wing patterns are different; hence it is desirable that the folding pattern in each wing is fixed for design and manufacturing. But that is not true, and the answer can be confirmed easily by continuous video recording of a single living rove beetle isolated from other individuals; you will see that the beetle folds its wings sometimes from the right side and sometimes from the left side.

Table 7.1 shows the frequency and order of folding directions obtained from continuous observation of five living *C. vestitus* specimens. In each specimen, approximately 20 consecutive wing-unfolding/wing-folding processes were investigated. It is difficult to identify any regularity in the results because the table shows all tendencies: right-dominant individuals, left-dominant individuals, and right-left-balanced individuals.

To clarify the reason for the two different folding modes is very difficult, but it may be speculated that this "right or left" option confers great advantages for expeditious wing folding and hiding in situations where obstructions prohibit movement of the abdomen. Another reason may be aerodynamical; fixing the folding direction would cause an asymmetric effect on the aerodynamic properties of wings, thus impairing flight.

Table 7.1 Results of continuous observations cf the folding direction in five specimens

	1	2	3	4	5	6	7	8	9	10	11	12	13	14	15	16	17	18	19	20	21	22	23	24	25	R : L
A	R	L	L	R	R	L	R	L	R	R	L	R	L	L	R	L	R	L	L							9 : 10
B	R	L	R	L	R	L	L	L	R	L	L	L	L	L	L	R	L	L	R	L						6 : 14
C	L	R	L	L	R	L	L	R	R	L	L	L	L	R	R	R	L	L	R	R	L	L	R	L	R	11 : 14
D	R	R	L	L	R	R	L	R	R	L	R	R	R	R	R	R	R	L	L	R	R	L				15 : 07
E	R	R	R	R	L	R	R	L	L	L	L	L	L	R	L	R	L	L	L	R						9 : 11

R: Right-overlying folding
L: Left-overlying folding

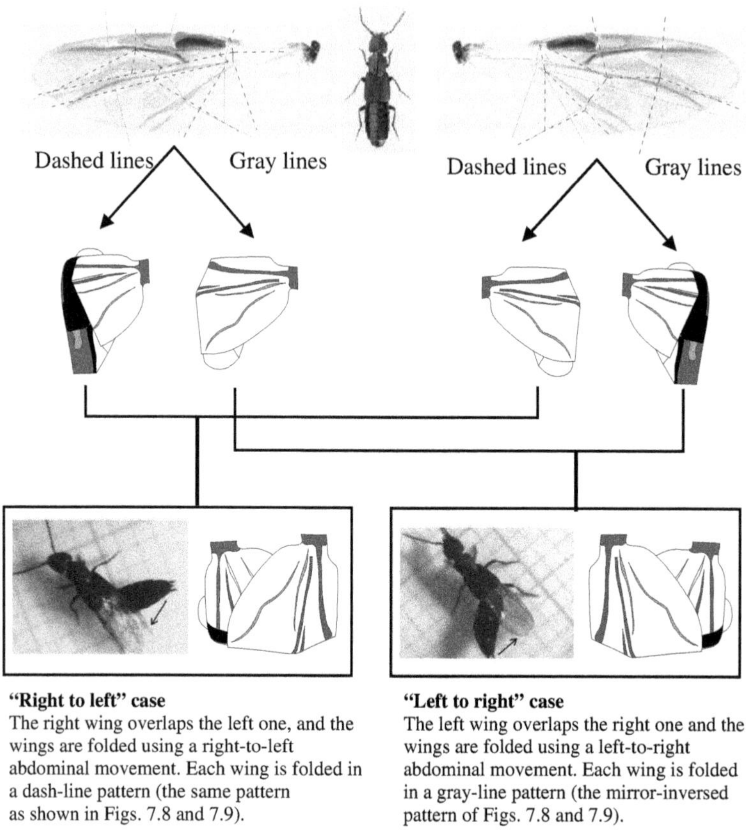

Figure 7.10 Two possible folding movements. Each individual rove beetle can fold its wings from both sides. Thus, each wing internalizes two crease patterns and can be folded in two different ways.

From an engineering perspective, what is so surprising is that one wing internalizes two crease patterns, allowing the structure to fold in two modes (Fig. 7.10). In other words, wings have three transforming forms: completely unfolded, right-overlying folding, and left-overlying folding. This fact defies the common wisdom that one deployable structure internalizes one folded shape. Though having such a complex transformation system, the wing itself seems to be very simple in terms of structure. If a human engineer tries to achieve this two-way-foldable wing by classical mechanisms, he or she will have to prepare many mechanical parts, including joints with rocking and release function, and the structure would be very complex. In contrast, veins in the wings of rove beetles have no clear hinges or bending points.

7.4 Conclusion

As shown in this chapter, rove beetles use quite a different method for wing actuation from that used in man-made machines. In a way, biomimetics is to learn the best solution that a species finds in the course of its evolution for various engineering problems. One of major advantages is that it enables top-down technological development instead of the usual bottom-up improvement of technology. Rove beetles use asymmetric patterns in folding symmetrically shaped structures. This is an off-the-wall idea that strays from the usual route of the research and development of deployable structures and therefore is expected to broaden the design possibilities for existing symmetrically shaped deployable structures, such as solar array paddles in satellites and wings of carrier-based aircraft. It is also expected that many articles of daily use, such as umbrellas and fans, which have not evolved any further, will be improved by the use of asymmetric folding patterns. The idea of a deployable structure that can be folded into two different shapes is also surprising. Although further investigations of wing structures are required, this capability will provide new possibilities for deployable structures and transformation mechanisms used in various machines.

Not only the wing transformation mechanism but also the rove beetle itself is an interesting research target as the model of an insect robot. It can run rapidly on the ground, crawl through small gaps, and move long distances by flying. This is the ideal model for a search robot that could search for signs of life trapped under wreckages resulting from disasters such as earthquakes and tsunamis. Deployable wings on these robots will be powerful tools enabling expansion of the search area.

Acknowledgments

The author thanks S. Yamamoto, M. Maruyama, and S. Nomura for giving entomological advice and Nobi-Tech., Ltd., and Keyence Corporation for providing high-speed cameras. This work was supported by a special fund from the Institute of Industrial Science, University of Tokyo; JSPS KAKENHI Grant Number 24860024; and the Asahi Glass Foundation Research Encouragement Grants (2016).

References

1. Danforth, B. N. and Michener, C. N. (1988). Wing folding in Hymenoptera, Ann. Entomol. Soc. Am., 81, pp. 342–349.
2. Wootton, R. J. (1995). Geometry and mechanics of insect hindwing fans: a modelling approach, Proc. R. Soc. B: Biol. Sci., 262, pp. 181–187.
3. Haas, F., Gorb, S. and Wootton, R. J. (2000). Elastic joints in Dermapteran hind wings: materials and wing folding, Arthropod Struct. Dev., 29, pp. 137–146.
4. Nojima, T. (2007). Origami modeling of functional structures based on organic pattern, Proc. VIPSI Tokyo.
5. Saito, K., Yamamoto, S., Maruyama, M. and Okabe, Y. (2014). Asymmetric hindwing folding in rove beetles, Proc. Natl. Acad. Sci. U. S. A., 111, pp. 16349–16352.
6. Forbes, W. T. M. (1924). How a beetle folds its wings, Psyche, 31, pp. 254–258.
7. Forbes, W. T. M. (1926). The wing folding pattern of the Coleoptera, J. N. Y. Entomol. Soc., 34, pp. 42–68.
8. Hammond, P. M. (1979). Carabid Beetles, Thrie Evolytion, Nautural History, and Classification, eds. Erwin, T. L., Ball, G. E. and Whitehead, D. R., Chapter 22; Wing-folding mechanisms of beetles, with special reference to investigations of adephagan phylogeny (Coleoptera), (Dr W. Junk, The Hague), pp. 113–180.
9. Brackenbury, J. H. (1994). Wing folding and free-flight kinematics in Coleoptera (Insecta): a comparative study, J. Zool. (Lond.), 232, pp. 253–283.
10. Haas, F. and Beutel, R. G. (2001). Wing folding and the functional morphology of the wing base in Coleoptera, Zoology, 104, pp. 123–141.
11. Sun, J., Ling, M., Wu, W., Bhushan, B. and Tong, J. (2014). The hydraulic mechanism of the unfolding of hind wings in Dorcus titanus platymelus (order: Coleoptera), Int. J. Mol. Sci., 15, pp. 6009–6018.
12. Kishimoto, N., et al. (2006). New deployable membrane structure models inspired by morphological changes in nature, Proc. 47th AIAA/ASME/ASCE/AHS/ASC Structures, Structural Dynamics, and Materials Conference, AIAA-2006-1898.

Chapter 8

Biotemplating Process for Electromagnetic Materials

Kaori Kamata
Chemistry Division, Department of Liberal Arts & Sciences, National Defense Medical College, 3-2 Namiki, Tokorozawa, Saitama 359-8513, Japan
ryu-te-n66@ndmc.ac.jp

8.1 Introduction

In recent years, the application range of electromagnetic-wave-absorbing material has dramatically expanded with the development of high-speed and large-capacity information communication technology. The targeting frequency band has reached up to the gigahertz/terahertz ranges, which have been attracting attention as new absorption bands. There are therefore increasing expectations for the high-frequency operation year by year [1, 2]. These technological trends are associated with the development of light sources, antennas, and detectors for the gigahertz/terahertz waves. More concrete applications, such as nondestructive testing, spectroscopic techniques, and effective electromagnetic wave

Industrial Biomimetics
Edited by Akihiro Miyauchi and Masatsugu Shimomura

ISBN 978-981-4800-07-5 (Hardcover), 978-0-429-05883-7 (eBook)
www.jennystanford.com

absorbers, corresponding to the high frequencies have been proposed and implemented. As electromagnetic wave absorbers for the millimeter-gigahertz band, magnetic materials such as ferrite have been regarded as promising candidates so far, and their high-frequency absorbers for the approximately 200 GHz band were reported with manufacturing methods and absorption mechanisms [3]. However, in a higher-frequency band, magnetization reversal of the magnetic material against the alternating magnetic field of the electromagnetic wave is an underlying problem to the operation principle itself. Additionally, the absorption efficiency decreases in a higher-frequency region, so a large amount of magnetic material dispersions becomes necessary. These problems still remain in engineering applications and are expected to be overcome, especially the problems related to lightweight thinning, visible light transparency, coating feasibility, and so on.

Conventionally, materials used for radio wave engineering are limited to metals such as copper and aluminum, ferrite, and carbon and insulating materials, such as resins [4], and working with material *chemistry* that can lead to the creation of a variety of material resources has not been well considered in material development for high-frequency engineering. In this chapter, the new research area relating to the fabrication process using biomaterials for high-frequency materials is introduced and summarized, along with their characteristics of electromagnetic wave response.

8.2 Electromagnetic Materials Utilizing Biomaterials

The operating principle of the electromagnetic wave absorber is roughly divided into conduction loss, dielectric loss, electromagnetic induction, and magnetic loss. By a combination of materials having these functions, the absorbers can be designed to work against a target frequency band. At present, the typical methods for preparing absorbers involve kneading a filler such as conductive fibers or a conductive powder that generates current under irradiation of electromagnetic waves, into urethane, or rubber, and processing the resulting composite filler into a designed concave-convex structure showing interference with or resonant responses to the wavelength of the targeting electromagnetic wave. Accordingly, it is possible to combine the conductive loss of filler and the dielectric loss of the dielectric relief structure, thereby realizing effective electromagnetic wave absorption. In this way, moderate conductivity and a periodic structure that interferes with electromagnetic waves are key for material designs. Therefore, utilization of biomaterials having a porous structure and/or a characteristic 3D microstructure has been proposed for the development of new electromagnetic-wave-responsive materials. We can see a wide variety of wonderful microstructures, for example, in the internal or entire structures of plants or algae. If conductivity can be imparted to various structures in such a natural world, it is possible to expect a higher-frequency response, essential to miniaturization of the feature size of the material structure, and cost reduction and mass production, which are difficult with the artifacts described above. As a part of this idea, biomimetics have been proposed and generated many functional materials and systems. Biomimetics and related studies have started to contribute to a wide variety of science and engineering fields with, for example, photonic crystals interfering with electromagnetic waves in the optical region [5], superhydrophobic materials [6], and high-efficiency adhesive tape [7].

On the other hand, the fabrication process directly uses natural resources. It has been reported that a composite sheet obtained by kneading porous carbon powder consisting of rice bran, rice hulls, or soybean hulls, originally having a porous internal structure, carbonized beforehand in high-temperature nitrogen gas, exhibits high shielding and excellent absorption properties in the radio frequency band of about 100 MHz [8]. This method is characterized by the use of biomaterials from agricultural waste rather than petroleum resources and can afford a new process for the fabrication of electromagnetic shielding/absorbing materials that retain their lightweight characteristic, high lubrication performance, and high electroconductivity, just like carbon materials. Meanwhile, structures with periodic or geometric features found in nature are also being utilized for the development of electromagnetic-wave-responsive materials. The fabrication process mainly consists of metal coating on surfaces of concavo-convex or helical nano-/microstructures found in nature. The resulting materials exhibit electromagnetic wave absorption characteristics based on interference or electromagnetic induction effect against incident electromagnetic waves, which have wavelengths comparable to the feature size of materials. Since the overall structure of a biomaterial is (molded) like a template with a conductive material, this fabrication method is called the biotemplate process.

The introduction of biomaterials having unique nano-/microstructures in electromagnetic wave applications was first attempted more than 15 years ago [9]. Initially, there were many reports simply focusing on the uniqueness of the ideas themselves, of which essential were the use of the overall structure of the biomaterial in the fabrication process. Few studies mentioned the essential characteristics and functions specific in the nano-/microstructures of materials. Table 8.1 lists typical studies in the early stage extended to the structure-specific functions with respect to the electromagnetic wave in order of the largest feature size of the material structure. The feature size is extensive, from nanometer (globular protein and linear DNA) to micrometer (plant spiral fiber). The targeting functions are plasmonics and photonic crystals based on interaction with light, as well as electromagnetic induction characteristics in radio wave regions of higher frequencies, and so on. Figure 8.1 indicates the biotemplated materials extracted from Table 8.1.

Table 8.1 Representative electromagnetic wave application research using biomaterials

Biomaterial	Coating material	Structure and function	Refs.
Plant fiber	Ag	Left-handed helix, electromagnetic microcoil	[10]
Rice hull	(Carbonization)	Nano-/microporous, electromagnetic absorber	[8]
Blue-green algae	Cu	Helical microstructure, electromagnetic microcoil	[11]
Diatom	Au	Plasmonics, surface-enhanced Raman scattering	[12]
Butterfly wing	Iron oxide, SiO_2	Bicontinuous gyroid, photonic crystals	[13, 14]

Biomaterial	Coating material	Structure and function	Refs.
Pollen	FeP	Hollow particle	[15]
Tobacco mosaic virus	SiO_2, iron oxide, quantum dots (CdS, PbS)	Nanotubes and their hierarchical organizations	[9]
Flagellated bacteria	SiO_2, TiO_2	Fiber and nanotube	[16, 17]
Insulin fibril	Au	Gold chain structure	[18]
Ferritin	GaAs, ZnO, Fe_3O_4	2D array of nanodisk, photonic reactor, memory	[19–21]
DNA	Au, Ag	Ag nanowire, DNA-linked Au network, sensing	[22]

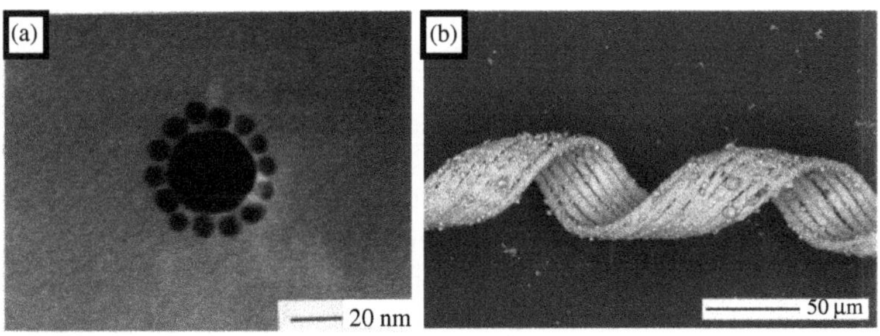

Figure 8.1 Examples of biotemplated materials: (A) a DNA-linked Au network [22] and (B) a silver microcoil templated with plant fiber (Reprinted from Ref. [10] with permission from John Wiley and Sons)

8.3 Application of the Biotemplate Process for the Fabrication of Electromagnetic Microcoils

The biggest feature of the biotemplate process is the ability to fabricate 3D microstructures that could not be artificially created. This is realized by borrowing an optimized structure through a long history of natural selection in the biological world. Additionally, the biotemplate is disposable and allows mass production through the fabrication process (Fig. 8.2). Accordingly, a biotemplating process using plant tissues with a unique spiral shape has been designed for the fabrication of an electromagnetic coil whose structural features are the same as those of the biotemplate [10].

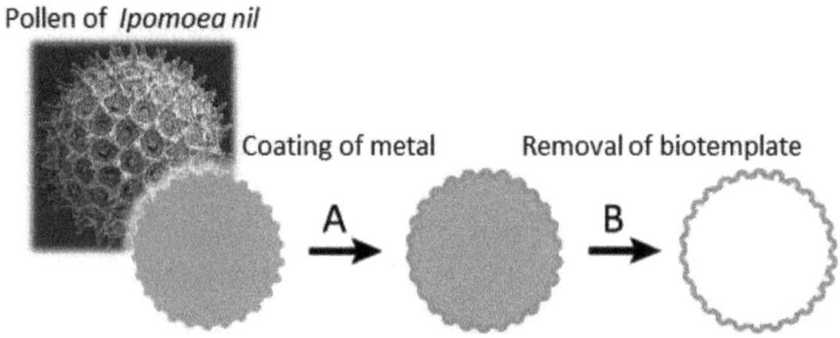

Figure 8.2 Schematic illustration of the biotemplating process using, for example, kinds of pollen. (A) Coating of a metal or targeting material onto the biotemplate surface and (B) removal of the biotemplate. The removal process is characteristic for biotemplates because the biotemplate is chemically decomposed and mainly not recycled, that is, it is disposable, unlike the expensive mold of nanoimprint.

The tissue of the vascular plant includes the xylem vessel, which is a tubular conductive tissue. The xylem vessel has microholes in its vessel wall, which forms a periodic pattern using a spiral, mesh, or ring structure. This vessel wall pattern is formed with a secondary cell wall consisting of dead cells after expanded growth and appears in the interior part of the stretchable tubular primary cell wall. The vessel wall pattern therefore contains crystalline cellulose as a main component, giving a supple and fine fiber with mechanical strength. Figure 8.3 shows a spiral vessel appearing from the fractured surface of a lotus root, which lets us see the perfect left-handed spiral microstructure. When the lotus rhizomes (lotus root) and its stems are cut, a number of spiral vessels appear as white fibers from the fractured surface. The spiral vessels of the lotus exhibit significantly higher mechanical strength rather than other tissues, which are immediately damaged and broken so that they are easily isolated from the vascular bundles [23, 24]. The spiral diameter is 50–80 μm, the pitch is 50–100 μm, and the length depends on the distance between the cutting surfaces, and it is possible to reach tens of centimeters. The lotus spiral vessels have a multiple spiral structure in which an average of seven single wires with diameters of about 5 μm are bundled. This vascular spiral pattern is completely left-handed, and its size varies depending on the plant species and parts, but the 3D spiral structure with a core space (air-core coil structure) can be found in either case. The design of electromagnetic materials using spiral vessels as the biotemplates is introduced below.

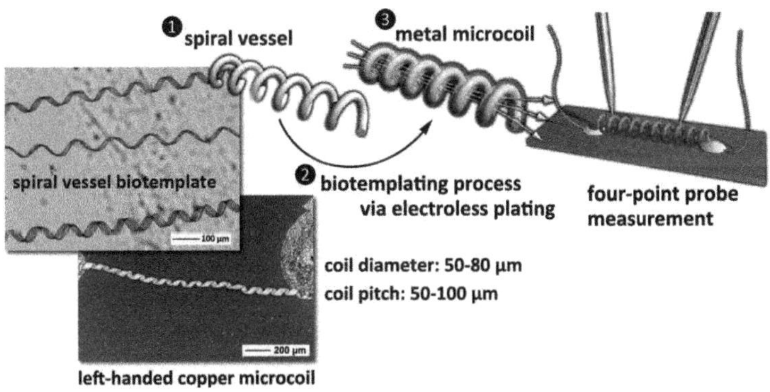

Figure 8.3 Biotemplating process on a spiral vessel: (1) extraction of a spiral vessel from plant tissue, (2) electroless plating, and (3) formation of a metal-deposited layer on the vessel surface. Electroless plating can be conducted with the spiral vessel even fixed on the substrate, which is accessible for successive electric and electromagnetic measurements. Two SEM images indicate the resulting metal microcoil fixed on the substrate with a silver paste

8.3.1 Electromagnetic Induction Properties of a Microcoil

Faraday discovered that electricity was flowing through the coil (electromagnetic induction) when the magnet was moved into and out of the air core of the coil. In that same period, Henry developed self-induction as a result of noticing sparks flying at the contacts when the current was turned off during the performance test of a telegraph machine that caused the bell to beat using an electromagnet. These two findings have helped develop many important technologies supporting current electronics (Fig. 8.4). In accordance with the change in magnetic flux generated by the alternating current applied to one coil, an adjacent coil generates the induced current. As described above, wireless communication technology working under the electromagnetic induction type makes use of "power transmission between disconnected and isolated coils." This electromagnetic wave propagating through the space between coils is a radio wave, as everyone knows. The frequency (f) of the circuit resonating with the electromagnetic wave is represented by $f = (2\pi\sqrt{LC})^{-1}$, where L is the inductance in Henry (H) and C is the capacitance in farad (F). For the coil, L is individually expressed by $L = (\mu_0\pi\ (D/2)^2\ N^2/l)\ K$, where D is the coil diameter, l is the coil length, N is the number of turns, μ_0 is vacuum magnetic permeability, and K is Nagaoka coefficient [25]. From these two equations expressing classical electromagnetism, it can be seen that the operating frequency becomes higher as the coil becomes smaller. In this manner, the microcoil obtained through the biotemplating process with a metal coating on the surface of the plant fiber (spiral shape at the micrometer level) can be logically expected to show the electromagnetic wave response in the higher-frequency region.

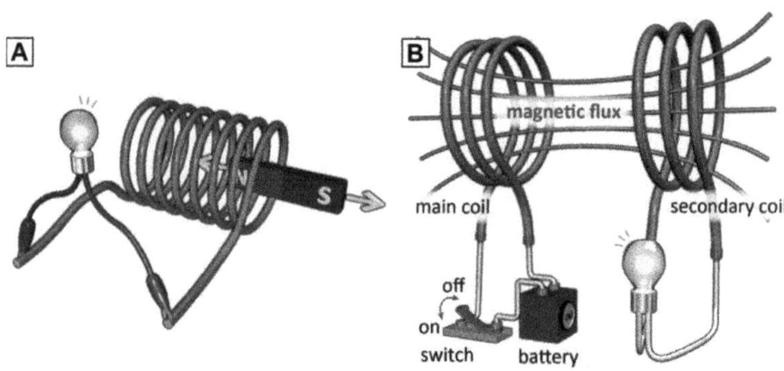

Figure 8.4 Simple experiments in electromagnetic induction. (A) When a permanent magnet is being moved into and out of a small coil, induced current flows in the coil. (B) Mutual induction occurs between two coils, where the main coil creates a magnetic flux as current flows and the other, secondary coil, has an electromotive force induced by the flux from the main coil.

8.3.2 Electromagnetic Properties of Lotus-Based Metallic Microcoils

The fabrication of a metallic microcoil employs a lotus spiral vessel as the biotemplate, a process that comprises isolation of the biotemplate by cutting the lotus rhizomes or stems, fixing of two rhizome pieces at an appropriate distance from each other (equal to the microcoil length) onto the substrate, and finally the coating of metal onto the surface of the biotemplate. Metal coating can be carried out by electroless plating, so the remaining process is to immerse the substrate with the biotemplate into the plating bath. By selection of the plating bath, it is possible to coat various kinds of metals onto the microcoils. Here, silver was used as the coating to create a highly electroconductive microcoil. From the scanning electron microscopy (SEM) image, it was found that the resulting silver microcoil has the same structure as the air-core coil that originated from the lotus spiral vessel. The electric resistance of the silver microcoil evaluated by the four-point probe method was proportional to the distance between the voltage probes, which follows Ohm's law. The electroconductivity reached the order of 10^6 S/cm, comparable to that of bulk silver. Further, the electromagnetic induction characteristics were examined to directly measure the inductance with the help of a superconducting quantum interference device (SQUID) [26]. The lotus-based silver microcoil fixed onto the substrate was inserted into the pickup loop coil of the SQUID, and current was applied from electric conduction wires connected to both ends of the microcoil. Theoretically, an induced current is generated through the pickup loop coil in response to a change in the magnetic flux density in the microcoil sample under current application. If magnetic susceptibility converted from the induced current of the inner pickup loop coil can be detected by a SQUID, the measurement setup proposed here is quite unique for detecting minute changes in the magnetic flux density generated through the microcoil. When a current between −5 and +5 mA was applied to a silver microcoil having a diameter of 60 μm, a length of 15 mm, and 29 turns at room temperature, the magnetic susceptibility was successfully detected as proportional to the applied current. The slope of the linear relationship between the magnetic susceptibility and the applied current gave the silver microcoil a minute inductance at the level of pico-Henry (25×10^{-12} H) (Fig. 8.5). For example, an inductor, capacitor, resistor circuit combining a silver microcoil having a very low inductance and a commercial capacitor with a capacitance of 0.1 pico-farad (0.1×10^{-12} F) will show a resonant frequency of about 100 GHz. From these studies, it is clear that the biotemplating process can be utilized for miniaturization of the microcoil, leading to an effectively high-frequency and broadband response.

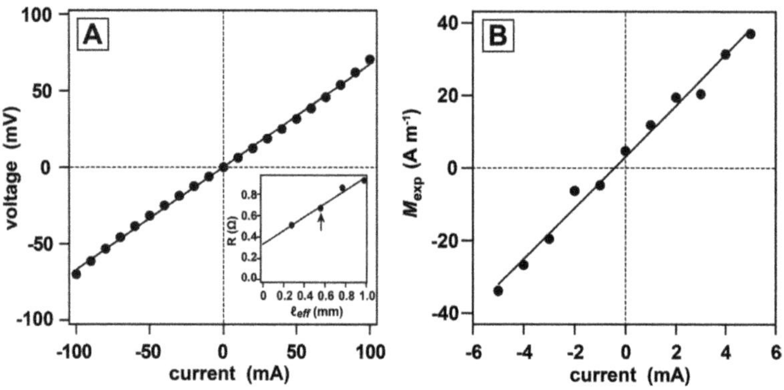

Figure 8.5 (A) *I-V* curve of the silver metal microcoil with a probe distance of 0.57 μm. The inset indicates a relationship between resistance and the probe distance, and the arrow shows the data at the *I-V* measurement. (B) Magnetization detected by a SQUID as a function of current passing through the microcoil.

8.4 Structure Control of a Metallic Microcoil via the Biotemplating Process and the Electromagnetic Characteristics

Structure control and alignment control of metallic microcoils have been expected to be two of the key features that will be integrated for new electromagnetic properties and optics such as metamaterials and chiral fields in the high-frequency region [27, 28]. Lithographic methods have been used to fabricate metallic nano- or microspirals as designed, while larger spirals, in the millimeter range, are made by classical winding or automatic wire bonder [29]. In view of mass fabrication, a wide variety of synthetic routes for spiral components have been proposed by chemical processes such as vapor deposition [30, 31] and supramolecular assembly [32]. The biotemplating process using microalgae is particularly effective because of proliferation by photosynthesis. This term introduces the function of a microalgae biotemplate for size-tunable microcoils and their electromagnetic properties.

Photosynthetic spiral microalgae, *Spirulina* (*Arthrospira platensis*), having a diameter of about 40 μm are easily grown, without the need for high technical skill for the cultivation. Furthermore, noteworthy here is that the spiral shape changes depending on the surrounding environment. By utilizing *Spirulina* as the biotemplate, a methodology for structural control of the biotemplated metallic microcoil can be designed simultaneously with mass production [33].

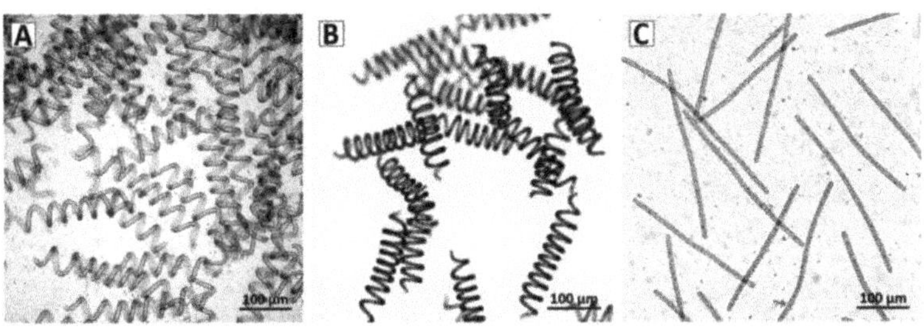

Figure 8.6 Optical micrographs of left-handed *Spirulina* compressed from the initial state (A) to state (B) under irradiation by light of a higher intensity and *Spirulina*'s linear strain (C) extended under mild conditions during cultivation for a long period.

One-handed *Spirulina* was cultured at 35°C in a society of toxicology medium [34] mainly composed of sodium bicarbonate. Under irradiation by light 3000 lx in intensity, the *Spirulina* gave a growth curve indicating a doubling in the number of *Spirulina* per day. For example, 500 *Spirulina* algae per 1 mL at the initial stage increased to 105 after one week of cultivation. The structural feature of the common *Spirulina* is a relatively loose winding of the left-handed spiral structure, which is usual in nature and also in the stock strain commercially available at a microbial collection center, such as the National Institute for Environmental Studies (NIES). The average pitch is 55 μm, and the pitch angle is around 34°. Here, it was found that an increase in the light intensity decreased the pitch and the pitch angle while keeping the other cultivation conditions as is. Fixing the tissue of the *Spirulina* in the compressing process during the cultivation yielded systematically different pitch and pitch angles, which resulted in the preparation of biotemplates with a wide variety of microcoil shapes (Fig. 8.6). On the other hand, when the successive cultivation was continued for longer than two months under light of an intensity lower than 3000 lx, the spiral structure completely disappeared and a linear-shaped strain was obtained. From the findings in the shape change depending on the culture environment, a method intentionally controlling the spiral structure of the *Spirulina* biotemplate was proposed, leading to the structural control of resulting metallic microcoils. The entire process is described in the following terms.

8.4.1 Tissue Fixation of the Spiral Biotemplate

A method for effectively maintaining the spiral structure controlled by the culture environment is required in order to utilize *Spirulina* as a biotemplate. The *Spirulina* surface is covered by a gelatinous extracellular membrane, which is easily removed at the step of metal coating or alcohol treatment [35], causing deformation of the precious spiral feature. The conventional technique of glutaraldehyde fixation [36] used for the immobilization of the biological sample is therefore effective in preserving the spiral feature for further biotemplating process to fabricate the metallic microcoil. The fixing reagent, glutaraldehyde, efficiently forms a cross-linked structure through a condensation reaction or an addition reaction with the amino terminuses of amino acids of proteins or peptides contained in the extracellular membrane. When the *Spirulina* is immersed overnight in a 3% glutaraldehyde phosphate buffer solution employed as the fixing solution here, the *Spirulina* tissue is cross-linked not only to the extracellular membrane but also to the inside of the *Spirulina* so that its spiral feature exhibits acid and alkaline tolerance. From a microscopic observation before and after the glutaraldehyde treatment, it can also be confirmed that the spiral structure is preserved (Fig. 8.7). Using the two-step process—structure control by the culture environment and tissue fixation—it became possible to prepare *Spirulina* biotemplates with systematically different spiral features, which can be stored for a long period.

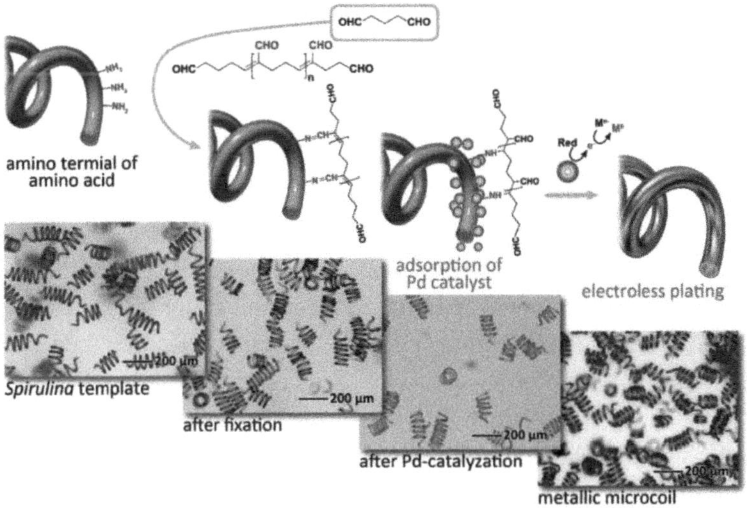

Figure 8.7 Biotemplating process using *Spirulina* in three steps: fixation of *Spirulina* tissue by glutaraldehyde, adsorption of Pd metal, and electroless plating. The template that appeared in the optical micrographs was a left-handed *Spirulina* strain (NIES-46).

8.4.2 Biotemplate Process Using the Fixed *Spirulina*

Spirulina is regarded as a fine spring because the spiral pitch systematically changes during the proliferation process. As in the case of the lotus-based microcoil mentioned above, fabrication of electromagnetic microcoils has been proposed that applies electroless plating to coat metal onto the surface of *Spirulina*. Electroless plating involves the following steps: (i) metallic clusters such as fine palladium particles are adsorbed along the surface through coordination bonding, (ii) the clusters work as a catalyst nucleus to oxidize the reducing agent in the plating bath, and (iii) metallization based on the reduction reaction of a metal ion occurs only in the vicinity of the catalyzed surface [37]. The fixed *Spirulina* surface is covered with many functional groups, such as amino acid residues and ester groups, being ligands effective for chemical adsorption of catalytic clusters. Therefore, even though the fixed *Spirulina* is dispersed in the solution, the catalyst nuclei are adsorbed only on the *Spirulina* surface and metal coating can occur exactly only on the surface, without free reduction of metal ions in the solution. Actually, metallic palladium adsorption on the *Spirulina* surface was confirmed after a simple preliminary experiment with a commercial plating pretreatment solution (OPC process AC) prepared for the adsorption of the palladium catalyst nuclei. Subsequently, electroless copper plating (OIC copper) was conducted and optical microscopy revealed that the resulting sample showed metallic luster based on the deposited copper (Fig. 8.7). The result also suggested that the coating of copper was completely formed along the surface of the *Spirulina*. Furthermore, no metallic particles precipitated in the plating bath. The thickness of the copper coating layer can be controlled in the range of 20 nm to 1 μm depending on the plating time and bath temperature. Using an electroless plating solution containing targeting metal ions, such as gold, silver, or nickel, as well as copper, various kinds of metal microcoils can be generated by the *Spirulina* templates.

8.4.3 Electromagnetic Property of a *Spirulina*-Templated Microcoil

When the thickness of a metal coating formed by electroless plating is adjusted to show electroconductivity comparable to that of a bulk metal, the microcoils function as electromagnetic coils. In addition, since *Spirulina* can provide biotemplates with various spiral features controllable in the cultivation process, *Spirulina*-templated microcoils fabricate metal microcoils having systematically different structural factors. The self-resonance wavelength (λ) of the electromagnetic coil is generally estimated to be in the range of $L/2 \leq \lambda \leq L$, where L is the circumference of the circular section in the coil [38]. In the case of a metallic microcoil fabricated by a *Spirulina* biotemplate, since L is approximately 150 μm, given that λ is 75 ~ 150 μm, the self-resonance frequency corresponds to 0.6 ~ 1.3 THz.

The response to the terahertz wave was performed by transmittance measurement of the sheet sample including the copper microcoil isotropically dispersed in paraffin transparent [39] in the terahertz band [40] (Fig. 8.8). The microcoil-dispersed paraffin sheet exhibited a very large transmission loss with a very low transmittance, 0.01% ~ 0.1% above 0.5 THz in frequency, while the copper microcoil concentration in the paraffin matrix was only 2 wt%. The reflectance separately measured was found to be 3%. The results of terahertz spectroscopies concluded that most of the incident terahertz wave was effectively absorbed by the microcoil sheet on the basis of the self-resonance of the copper microcoil (Fig. 8.8 middle) [36]. As a function specific to the spiral feature, a unique response was shown by the microcoil sheets against the terahertz circular polarization, that is, the microcoil selectively absorbed the circular polarization with the same polarization direction as the handedness of the microcoil (Fig. 8.8 right). Note that a recessive right-handed *Spirulina* biotemplate was prepared from the mutant of the right-handed strain, which is dominant in nature, to examine the effect of handedness on polarization property (Fig. 8.9). The *Spirulina*-based biotemplating process can contribute to the development of electromagnetic microcoils showing structure-specific functions as well as their mass production [11]. There are two outstanding issues in the *Spirulina* biotemplating process, as follows: A technique of down-sizing of a *Spirulina* biotemplate is meeting the increasing requirement for the higher-frequency response in the, for example, far-infrared or infrared region. The second issue is the orientation control of microcoils in the sheet, leading to infinite possibilities for a highly sensitive circular polarization response [41].

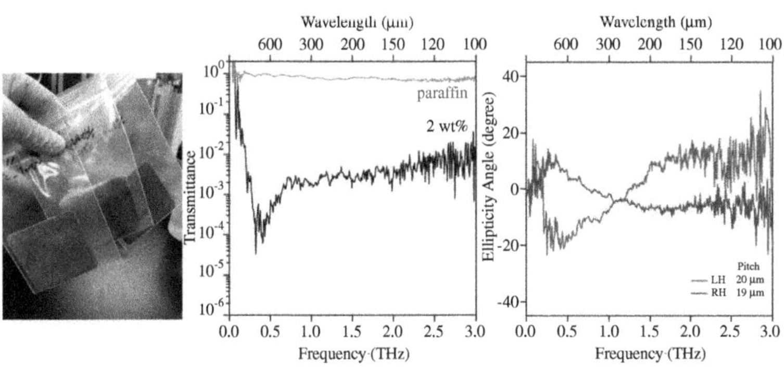

Figure 8.8 A paraffin dispersion sheet with 2 wt% of left-handed copper microcoils (left) and its transmission spectrum in the range of THz wave (middle), and spectra of ellipticity angles (right) of paraffin sheets also including 2 wt% of left- and right-handed microcoils.

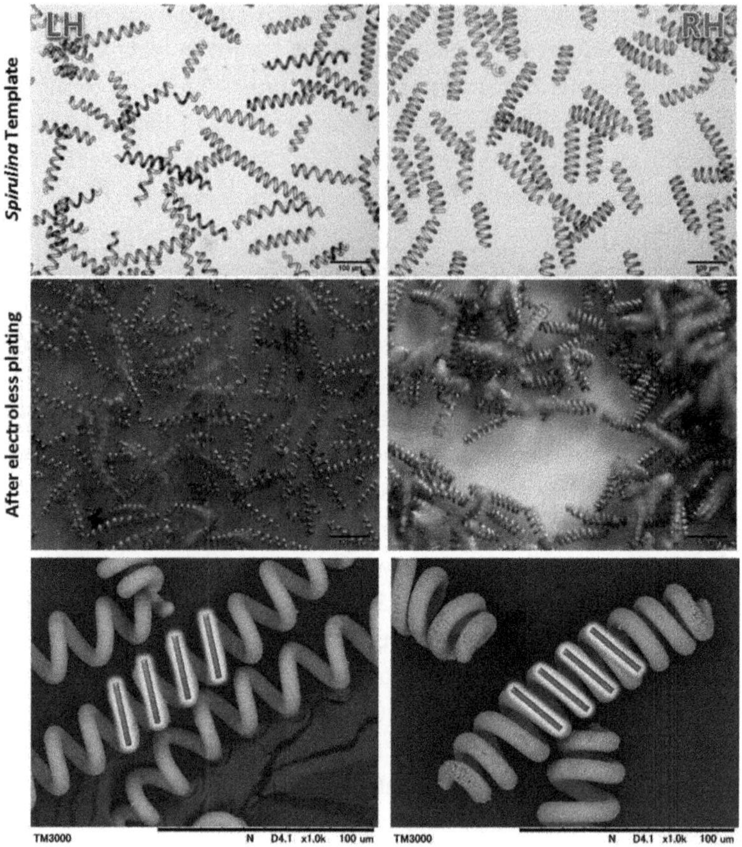

Figure 8.9 An enantiomeric pair of *Spirulina* templates (top) generated the corresponding copper microcoils: (left column) left-handed and (right column) right-handed microcoils.

8.5 Conclusion

Biomaterials are wealthy material resources due to a variety of chemical constituents; their complex 3D structures, superior to the leading edge of conventional manufacturing technology and the diversity of these structures [42]. The biotemplating process, taking advantage of the outstanding characteristics of the biomaterials, can generate various kinds of metallic microstructures with their structure-specific functions. The microcoils introduced herein can freely tune their operative frequencies by controlling their own structures, which has never been achieved in conventional systems. Effectively preparing the dispersion sheet makes it possible to develop advanced filters, such as low-pass, high-pass, bandpass, and band-stop filters, operating at any frequency, even in the wide range of frequencies. Extending the targets of biomaterials as biotemplates can improve the performance and variations for future technology in electromagnetic wave applications [43].

References

1. Tonouchi, M. (2007). Cutting-edge terahertz technology, Nat. Photonics, 1, pp. 97–105.
2. Dhillon, S. S., Vitiello, M. S., Linfield, E. H., Davies, A. G., Hoffmann, M. C., Booske, J., Paoloni, C., Gensch, M., Weightman, P., Williams, G. P., Castro-Camus, E., Cumming, D. R. S., Simoens, F., Escorcia-Carranza, I., Grant, J., Lucyszyn, S., Kuwata-Gonokami, M., Konishi, K., Koch, M., Schmuttenmaer, C. A., Cocker, T. L., Huber, R., Markelz, A. G., Taylor, Z. D., Wallace, J., Axel Zeitler, V. P., Sibik, J., Korter, T. M., Ellison, B., Rea, S., Goldsmith, P., Cooper, K. B., Appleby, R., Pardo, D., Huggard, P. G., Krozer, V., Shams, H., Fice, M., Renaud, C., Seeds, A., Stöhr, A., Naftaly, M., Ridler, N., Clarke, R., Cunningham, J. E. and Johnston, M. B. (2017). The 2017 terahertz science and technology roadmap, J. Phys. D: Appl. Phys., 50, p. 043001.
3. Namai, A., Yoshikiyo, M., Yamada, K., Sakurai, S., Goto, T., Yoshida, T., Miyazaki, T., Nakajima, M., Suemoto, T., Tokoro, H. and Ohkoshi, S. (2012). Hard magnetic ferrite with a gigantic coercivity and high frequency millimetre wave rotation, Nat. Commun., 3, p. 1035.
4. Petrov, V. M. and Gagulin, V. V. (2001). Microwave absorbing materials, Inorg. Mater., 37, pp. 93–98.
5. Mann, S. (1993). Molecular tectonics in biomineralization and biomimetic materials chemistry, Nature, 365, pp. 499–505.
6. Sun, T., Feng, L., Gao, X. and Jiang, L. (2005). Bioinspired surfaces with special wettability, Acc. Chem. Res., 38, pp. 644–652.
7. Hosoda, N. (2013). Handbook of Sustainable Engineering, eds. Kauffman, J. and Lee, K.-M., Section VII New Materials; Mechanisms of organisms as environment-friendly materials design tools (Springer), pp. 1249–1261.

8. Kaneta, M., Takahashi, T., Kanaya, T. and Iizuka, H. (2009). Electromagnetic wave absorption by rice hull carbon, Trans. Mater. Res. Soc. Jpn., 34, pp. 651–654.

9. Shenton, W., Douglas, T., Young, M., Stubbs, G. and Mann, S. (1999). Inorganic–organic nanotube composites from template mineralization of tobacco Mosaic virus, Adv. Mater., 11, pp. 253–256.

10. Kamata, K., Suzuki, S., Ohtsuka, M., Nakagawa, M., Iyoda, T. and Yamada, A. (2011). Fabrication of left-handed metal microcoil from spiral vessel of vascular plant, Adv. Mater., 23, pp. 5509–5513.

11. Kamata, K., Piao, Z., Suzuki, S., Fujimori, T., Tajiri, W., Nagai,K., Iyoda, T., Yamada, A., Hayakawa, T., Ishikawa, M., Horaguchi, S., Belay, A. M., Tanaka, T., Takano, K. and Hangyo, M. (2014). Spirulina-templated metal microcoils with controlled helical structures for THz electromagnetic responses, Sci. Rep., 4, p. 4919.

12. Rosi, N. L., Thaxton, C. S. and Mirkin, C. A. (2004). Control of nanoparticle assembly by using DNA-modified diatom templates, Angew. Chem. Int. Ed., 43, pp. 5500–5503.

13. Peng, W., Zhu, S., Wang, W., Zhang, W., Gu, J., Hu, X., Zhang, D. and Chen, Z. (2012). 3D network magnetophotonic crystals fabricated on morpho butterfly wing templates, Adv. Funct. Mater., 22, pp. 2072–2080.

14. Mille, C., Tyrode, E. C. and Corkery, R. W. (2011). Inorganic chiral 3-D photonic crystals with bicontinuous gyroid structure replicated from butterfly wing scales, Chem. Commun., 47, pp. 9873–9875.

15. Cao, F. and Li, D. (2010). Biotemplate synthesis of monodispersed iron phosphate hollow microspheres, Bioinsp. Biomim., 5, pp. 016005–016010.

16. Jo, W., Freedman, K. J., Yi, D. K. and Kim, M. J. (2012). Fabrication of tunable silica-mineralized nanotubes using flagella as bio-templates, Nanotechnology, 23, pp. 055601–055606.

17. Li, D., Mathew, B. and Mao, C. (2012). Biotemplated synthesis of hollow double-layered core/shell titania/silica nanotubes under ambient conditions, Small, 8, pp. 3691–3697.

18. Hsieh, S., Hsieh, C.-W. (2010). Alignment of gold nanoparticles using insulin fibrils as a sacrificial biotemplate, Chem. Commun., 46, pp. 7355–7357.

19. Wang, X.-Y., Huang C.-H., Tsukamoto, R., Mortemousque, P. A., Itoh, K. M., Ohno, Y. and Samukawa, S. (2011). Damage-free top-down processes for fabricating two-dimensional arrays of 7 nm GaAs nanodiscs using

bio-templates and neutral beam etching, Nanotechnology, 22, p. 365301.

20. Uenuma, M., Ban, T., Okamoto, N., Zheng, B., Kakihara, Y., Horita, M., Ishikawa, Y., Yamashita, I. and Uraoka, Y. (2013). Memristive nanoparticles formed using a biotemplate, RSC Adv., 3, pp. 18044–18048.

21. Suzumoto, Y., Okada, M. and Yamashita, I. (2012). Fabrication of zinc oxide semiconductor nanoparticles in the apoferritin cavity, Cryst. Growth Des., 12, pp. 4130–4134.

22. Storhoff, J. J. and Mirkin, C. A. (1999). Programmed materials synthesis with DNA, Chem. Rev., 99, pp. 1849–1862.

23. Ogasawara, S. (2005). Lotus fiber—comparison with the lotus fiber weaving at Myanmar and the picture scroll of Taima-mandala, Fiber J. (Sen'i Gakkaishi), 61, pp. 298–302 (in Japanese).

24. Kamata, K., Akimoto, Y. and Iyoda, T. (2013). Fabrication of metal microcoil from plant tissue, Cellul. Commun., 20, pp. 69–73 (in Japanese).

25. Nagaoka, H. (1909). The inductance coefficients of solenoids, J. Coll. Sci. Tokyo, 27, pp. 1–33.

26. Kamata, K., Ohtsuka, M., Nakagawa, M. and Iyoda, T. (2008). Expected materials for the future, Mirai Zairyo, 8, pp. 2–4 (in Japanese).

27. Pendry, J. B. (2004). A chiral route to negative refraction, Science, 306, pp. 1353–1355.

28. Gansel, J. K., Thiel, M., Rill, M. S., Decker, M., Bade, K., Saile, V., Freymann, G., Linden, S. and Wegener, M. (2009). Gold helix photonic metamaterial as broadband circular polarizer, Science, 325, pp. 1513–1515.

29. Kratt, K., Badilita, V., Burger, T., Korvink, J. G. and Wallrabe, U. (2010). A fully MEMS-compatible process for 3D high aspect ratio micro coils obtained with an automatic wire bonder, J. Micromech. Microeng., 20, p. 015021.

30. Hwang, G., Hashimoto, H., Bell, D. J., Dong, L., Nelson, B. J. and Schön, S. (2009). Piezoresistive InGaAs/GaAs nanosprings with metal connectors, Nano Lett., 9, pp. 554–561.

31. Motojima, S. and Chen, X. (2007). Preparation and characterization of carbon microcoils (CMCs), Bull. Chem. Soc. Jpn., 80, pp. 449–455.

32. Shimizu, T., Masuda, M., Minamikawa, H. (2005). Supramolecular nanotube architectures based on amphiphilic molecules, Chem. Rev., 105, pp. 1401–1444.

33. Iyoda, T., Kamata, K. and Yamada, A. (2009). Helical fine structure, method for producing the same, and electric-wave shield or absorber using the helical fine structure, PCT/JP2010/057306.

34. Ogawa, T. and Terui, G. (1972). Growth kinetics of Spirulina platensis in autotrophic and mixotrophic cultures, Proceedings of IV IFS: Fermentation Technology Today, Society of Fermentation Technology, Osaka, pp. 543–549.

35. Tada, T. and Tanaka, H. (2010). Shokubutsu no Seitai Zukan, Zoho Kaitei: Dai Shizen no Fushigi (Gakken, Tokyo) (in Japanese).

36. Migneault, I., Dartiguenave, C., Bertrand, M. J. and Waldron, K. C. (2004). Glutaraldehyde: behavior in aqueous solution, reaction with proteins, and application to enzyme crosslinking, BioTechniques, 37, pp. 790–802.

37. Bard, A. J. and Faulkner, L. R. (2000). Electrochemical Methods: Fundamentals and Applications (John Wiley & Sons).

38. Balanis, C. A. (2005). Antenna Theory: Analysis and Design (John Wiley & Sons).

39. Enatsu, T., Kitahara, H., Takano, K., Nagashima, T., Tani, M., Hangyo, M., Miura, Y. and Sawai, T. (2007). Terahertz spectroscopic imaging of paraffin-embedded liver cancer samples, 2007 Joint 32nd International Conference on Infrared and Millimeter Waves and the 15th International Conference on Terahertz Electronics, pp. 557–558.

40. Practically, the transmission loss includes the energy dissipation process by Joule heat generated from the induced current through the self-resonance of coil as well as the Rayleigh scattering to the outside of optical path connecting the light source and the detector. The estimation of both contributions is currently under consideration, but the angle dependent measurement and the thermal measurement effective for the estimation are difficult to directly measure in the terahertz region. Here, we described "absorption" including every factors mentioned above.

41. Ogino, S., Iyoda, T., Kamata, K. and Tajiri, W. (2015). Electromagnetic wave absorber, JP 2015-95638.

42. Shimomura, M. (2012). Engineering Biomimetics: Integration of Biology and Nanotechnology (Springer).

43. Kamata, K., Park, T. and Iyoda, T. (2014). Micro metal structure, PCT/JP2015/061796, WO 2015166816 A1.

Chapter 9

Application of Structural Color

Haruko Hirose
Teijin Limited, 4-3-2 Asahi-ga-oka, Hino-shi, Tokyo 191-8512, Japan
h.hirose@teijin.co.jp

Lots of new products are now being developing using the mechanism of a morpho butterfly wing, which is one of the good examples of biomimetics research [1].

In this chapter, the structure and the coloring mechanism of a morpho butterfly wing are described. Then structural color fibers and films that are developed using them are introduced.

9.1 Structural Color

A chromogenic effect is caused when a white light is projected on a fine structure of a size equal to or shorter than a light wavelength—it is called coloring. The colors existing in nature are roughly categorized into two types: one is "structural color," which is caused by this phenomenon, and the other is "pigment color."

Industrial Biomimetics
Edited by Akihiro Miyauchi and Masatsugu Shimomura

ISBN 978-981-4800-07-5 (Hardcover), 978-0-429-05883-7 (eBook)
www.jennystanford.com

Structural color is caused by the scattering, refraction, interference, diffraction, and so forth, of light. And the structures that cause structural coloring include the multilayer structure that is formed by laminating thin plates with mutually different refractive indexes.

In nature, many structural colors caused by the interference of light are already recognized. For example, the colors of a butterfly, a scarabaeid beetle, a pigeon's neck, a pearl, an abalone's shell, a squid's reflector, an opal, and a jewel beetle (Fig. 9.1) fall under the category of structural colors [2].

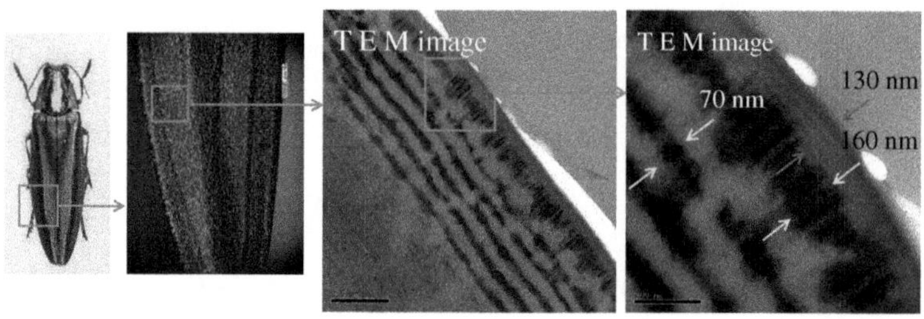

Figure 9.1 *Chrysochroa fulgidissima* wing and TEM images of the epidermis.

9.2 Structure of the Morpho Butterfly Wing

A male morpho butterfly's wing is decorated by scales, and the dimension of one scale is approximately 50 μm (width) × 200 μm (length) × 4 μm (thickness). The scale has many minute ridges, and the cross section of a ridge consists of layered structures of lamellae. The lamellae are made of a protein with a refractive index of 1.4–1.5 and a width of 70–80 nm. Between lamellae, there are air layers with a refractive index of 1 and a width of 140–160 nm. And a ridge has about 17 layers of both lamellae and air (Fig. 9.2). Furthermore, the thickness of the lower part of the ridge is 540 nm and the thickness of the upper part of the ridge is 50 nm, which makes it easy to allow the sunshine to reach the bottom of the ridge. In addition, the heights of the neighboring ridges are irregular. As a result, the wing is recognized as blue in color when viewed from any direction [3].

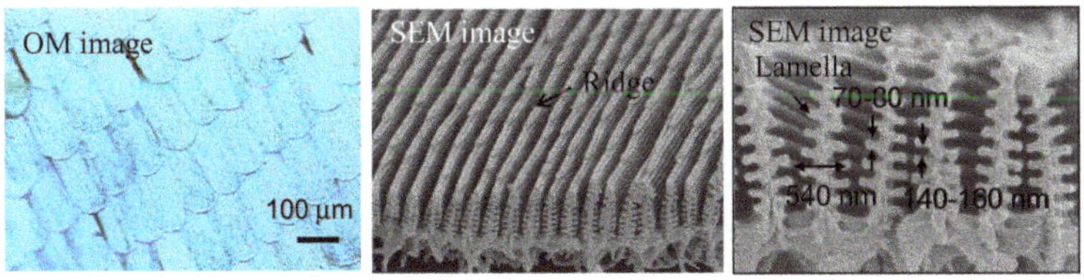

Figure 9.2 Wing of the morpho butterfly and SEM images of a scale.

9.3 Mechanisms of Structural Color

In physics, the property of overlapping of multiple waves, a new wave amplifying and multiple waves negating mutually, is called interference. There are two types of interference, waves strengthening and weakening mutually. As light passes through the thin layer, light interference occurs, and it follows that specific wavelengths of light are strengthened and the other specific wavelengths of light are weakened. After that, light is incident on the eye (Fig. 9.3). The thin film interference theory shown in Fig. 9.4 describes that a reflection wavelength λ (the color recognized by an eye) and a reflectance R (light intensity made incident on an eye) are shown by the following equations:

$$\lambda = 2(n_1 d_1 \cos\theta_1 + n_2 d_2 \cos\theta_2) \tag{9.1}$$

$$R = (n_1 - n_1)^2/(n_1 + n_2)^2 \tag{9.2}$$

Here, n is the refractive index, d is the thickness of the layer, and θ is the angle of refraction.

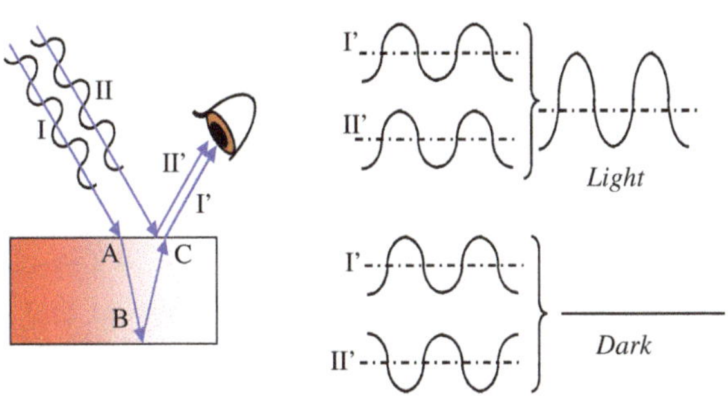

Figure 9.3 Mechanism of interference.

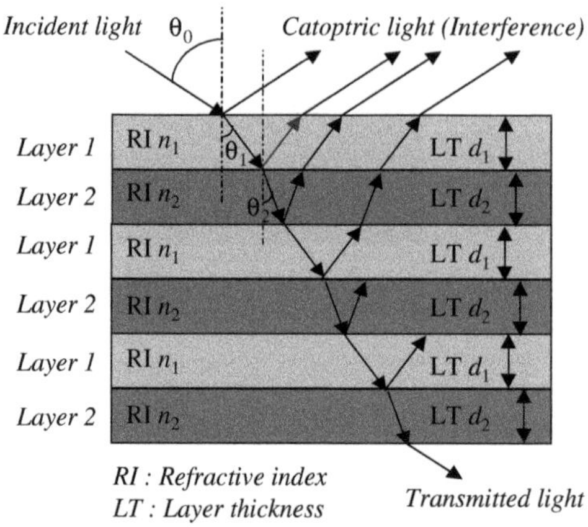

Figure 9.4 Schematic diagram of interference in a multilayer structure.

The larger the refractive index difference between the two polymer layers, the higher the reflectance. This means that the larger the refractive index difference between the two polymer layers, the brighter the color. That is, the interference coloring effects is maximized when $n_1d_1 = n_2d_2 = \lambda/4$ [4].

Multilayer wiring structures of a protein layer of about 70 nm and an air layer of about 140 nm are formed in the morpho butterfly wing. The reflected wavelength becomes about 490 nm at $\theta = 0$, and the morpho butterfly wing reflects blue light.

9.4 Development of the Structural Color Fiber "MORPHOTEX®"

It was predicted that structural color could be made by the combination of two kinds of polymers having different refractive indexes. But there were two major problems, stable color representation and control of the layered structures at the nanolevel.

To select adequate polymers, the following conditions had to be considered: a large difference in the refractive indexes of two polymers, easy processing of fiber, high strength of and excellent durability as a fiber, etc. After close examination of polymers, polyethylene terephthalate (PET), having a high refractive index, and nylon, having a low refractive index, were selected (Table 9.1).

Table 9.1 Physical properties of the polymer

High-refractive-index polymer

Polymer	Refractive Index	Δn	Tg(°C)	Tm(°C)
PEN	1.63	0.49	121	270
PC	1.59	0.20	145	280
PS	1.59	0.19	100	240
PET	1.58	0.22	69	260

Low-refractive-index polymer

Polymer	Refractive Index	Δn	Tg(°C)	Tm(°C)
Ny	1.53	0.08	40	260
PMMA	1.49	~0	105	180
Polymethylpentene	1.46	-	80	240
Fluorinated PMMA	1.40	-	-	160

But the difference in the refractive indexes of PET and nylon is only 0.05. This value is too small compared with the case of the morpho butterfly wing, where the difference in the refractive indexes is 0.5. To solve this problem, the number of layers was increased to 61, the cross-sectional shape of the fiber was made into a flat one, and the thickness of the layer was controlled within a fixed range, 70–100 nm (Fig. 9.5).

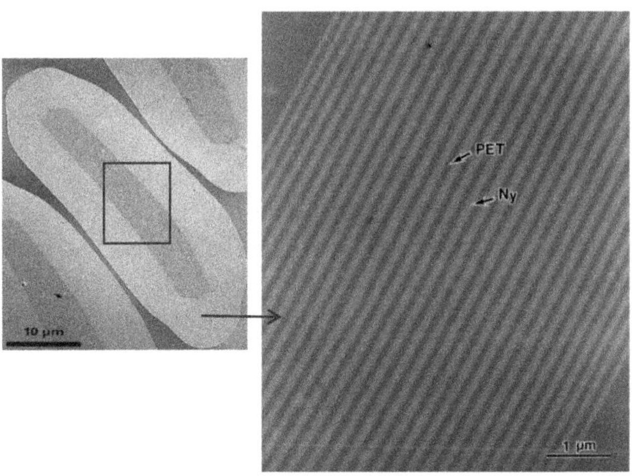

Figure 9.5 TEM images of a light interference fiber section.

These countermeasures made it possible to successfully make structural color fibers. Furthermore, it was possible to create three tones of colors (blue, green, and red) by changing the thickness of the layer, and stable coloring quality could be realized with a tensile strength of 3–4 cN/dtex and dry heat shrinkage of under 4% at 150°C. In addition, the optimized design of the spinneret and development of the direct stretching method have allowed the commercial production of the structural color fiber [5, 6].

MORPHOTEX® is a new fiber jointly developed by Nissan Motor Co., Ltd., Tanaka Kikinzoku Kogyo K.K., and Teijin Limited. In other words, MORPHOTEX® has been developed by the crystallization of the core technologies of the three companies in a cross-field partnership.

Made of colors created by light interference, MORPHOTEX® which colors are made by the light interference has some wonderful features. It has a clearly shining luster and changes its color when seen from different angles. Its color does not fade or change due to the structural coloring, and it is environment friendly because it is not dyed. Using these features, MORPHOTEX® has been applied in various fields. For example, as a filament, MORPHOTEX® is used in the fields of ladies' garments, sportswear, bags, shoes, curtains, interior materials, and so forth. On the other hand, as a staple fiber, MORPHOTEX® is used in the fields of coating automotive bodies and interior parts, cosmetics, and so forth.

9.5 Development of the Light Interference Film "Multilayer Film"

Resins are used for external materials in automobiles, electronic appliances, and electric products because resins are suitable for the 3D-shaped molding method that enables lightweight and intricately shaped molding products. However, in recent years, there has been a demand for an improvement in the design of molding products in order to make a difference. For this purpose, multilayer film (MLF), having structural coloring function, was developed by using the interference effect. MLF is a laminated structure made of alternate stacking of more than a few hundred layers composed of two resin layers with different refractive indexes, and the thickness between the laminated layers is under 120 nm (Fig. 9.6).

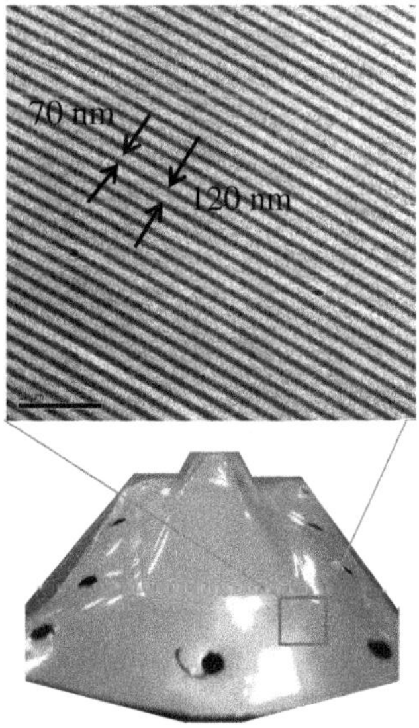

Figure 9.6 Three-dimensional molding and TEM image of MLF.

Two types of resins are individually melted using a different extruder. After being laminated in a melted state, the resins are extruded as a multilayer sheet using the multilayer laminated die block. Then, the sheet is quickly cooled on a quench drum. After quenching, the sheet is extended by sequential biaxial stretching so as to let the drawing directions cross at right angles to each other at a temperature higher than the glass-transition temperature of the resins. Finally, the film is heat-set in an oven (Fig. 9.7).

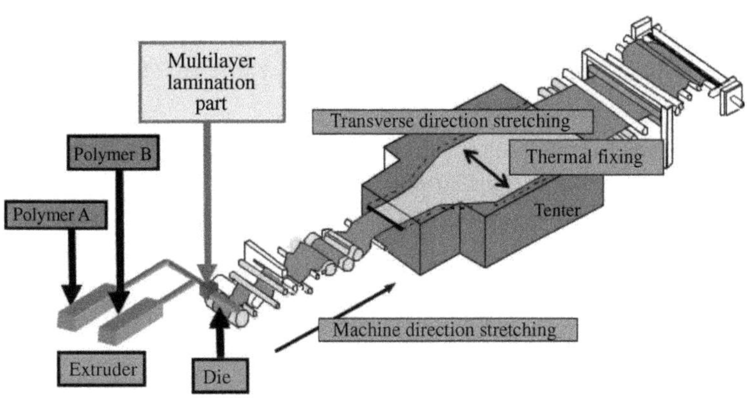

Figure 9.7 Manufacturing process of MLF.

By changing the thickness between layers, you can successfully create reflected colors that are blue, green, or red. The larger the refractive index difference between two resins or the larger the number of laminated layers, the higher the reflectance of light, and the color of the film can look dark and deep.

In addition, the thickness of the film can be varied from place to place by using insert molding. Furthermore, it is possible to combine these films using an updated printing technology. And this makes it possible to express a particular design [7].

In the case of structural color, while the structure is maintained, there is no discoloration and deterioration, and the color is maintained forever. And the production process has a low impact on the environment because the process does not require things like a pigment, a dye, or a large volume of water.

We will continue to develop new biomimetics products that have a low impact on the environment by collaboration with entities in other fields in the future.

References

1. Hirose, H. (2011). Development of light interference fiber and super multilayered film mimicking a morph butterfly, *Kobunshi*, **60**(5), p. 298.
2. Shimomura, M. (2012). Front of next generation of biomimetics research (CMC Publishing), pp. 288–291 (in Japanese).
3. Tabata, H., et al. (1996). Microstructures and optical properties of scales of butterfly wings, *Opt. Rev.*, **3**(2), pp. 139–145.
4. Radford, J. A., et al. (1973). Reflectivity of iridescent coextruded multilayered plastic films, *Polym. Eng. Sci.*, **13**(3), pp. 216–221.
5. Japan Patent 2890984 B2.
6. Yoshimura, M. (2003). Structurally colored fiber "MORPHOTE®", *Kobunshi*, **52**(11), pp. 826–828 (in Japanese).
7. Oyamatsu, A. (2011). *Plastic Age*, pp. 54–56 (in Japanese).

Chapter 10

Moth Eye–Type Antireflection Films

Yoshihiro Uozu
Mitsubishi Chemical Co., Ltd., 10-1 Daikoku-cho, Tsurumi-ku, Yokohama, Kanagawa 230-0053, Japan
uozu.yoshihiro.ma@m-chemical.co.jp

Moth eye surfaces can prevent reflection by changing the refractive index continuously between air and the matrices, which theoretically was already known. The technology to form such structures when small areas are involved has been developed, but it is difficult to form moth eye surfaces when large areas are involved. Recently, continuous photonanoimprinting of moth eye structures has been developed with a seamless anodic porous alumina mold. This chapter explores the technology of this development.

10.1 Introduction

With progress in the information-oriented society, there is a lot of digital information at play in the world. In the world of the Internet, a broadband connection has become a standard. The TV and the PC have become universal, and quite a few people have come to possess

Industrial Biomimetics
Edited by Akihiro Miyauchi and Masatsugu Shimomura

ISBN 978-981-4800-07-5 (Hardcover), 978-0-429-05883-7 (eBook)
www.jennystanford.com

more than one device. Furthermore, the number of cell phones and smartphones exceeds the number of people across nations. As fusion of the smartphone and the TV came about with the start of the one-segment television, a demand arose for smartphones to have displays of the same resolution as TVs have. In addition, the number of car navigation systems increased, and we can enjoy entertainment in the car. On streets and in trains, advertisements and information using display boards have become common. It's clear that displays have become the norm across the world. And higher-resolution, higher-definition displays are on the way.

The problem is that reflected light makes images less visible. To improve the display characteristics, reducing the effect of the reflected lights has become important. The most effective means of reducing the effects of reflected light are moth eye antireflection (AR) films.

10.2 Review of Reflection-Reducing Films

Generally, commercial reflection-reducing films are of two types, AR films and antiglare (AG) films. An AR film reduces the reflection itself to improve the image quality. An AG film reduces the reflection by scattering the light. These days, as flat panel displays (FPDs) become larger, there is a need to improve image quality. For an AG film, picture clarity needs to be improved by removing the whitish tinge. For an AR film, the reflection needs to be reduced. An extremely low reflectance film helps address both requirements. Most of the currently marketed AR films are multilayer type. The multilayer type AR films can improve a characteristic by increasing the layer numbers [1]. As FPDs should basically be low in cost, two-layer-type films are usually used.

10.3 Moth Eye Surface

10.3.1 Natural Moth Eye Surface

Figure 10.1 is a photograph of a natural moth eye structure. The moth eye has an unevenness to the surface to avoid reflection.

The structure serves as a survival mechanism by making the animal less vulnerable to predators [2]. Recently, research and development has been promoted for the industrial production of AR moth eye structures. This research and development is one of the representatives in biomimetics.

Figure 10.1 Natural moth eye structure.

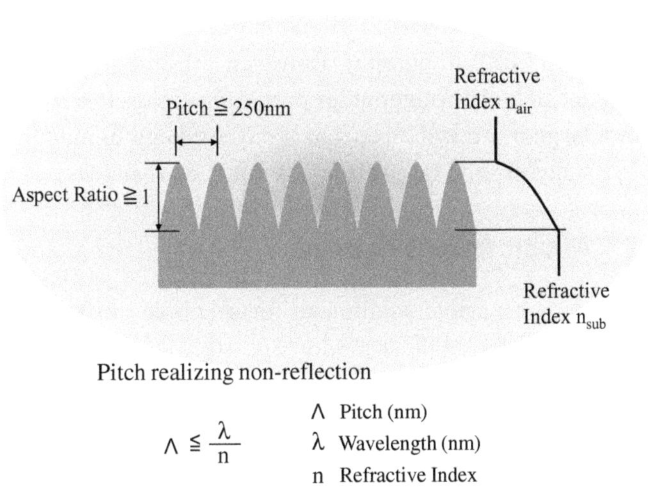

Figure 10.2 Image of our artificial moth eye surface.

10.3.2 Artificial Moth Eye Surface

Figure 10.2 shows the image of our artificial moth eye surface. The moth eye surface has a minute unevenness on the submicron order. There is a decrease in the refractive index from 1.5 on a plastic substrate to 1 on a surface that is the value of air. The difference in the refractive indexes is about 0.5. Individual projections have a cone-like shape. The diameter is lower than 250 nm, and the aspect ratio is larger than 1 [3]. When the pitch between adjacent projections is lower than the value of the wavelength of the incident light divided by the refractive index of the substrate, the reflection becomes negligible. If the wavelength is 400 nm and the refractive index is 1.5, the pitch is required to be lower than 267 nm.

10.4 Production Process of the Moth Eye Antireflection Film

10.4.1 Conventional Process

The moth eye structure can be made on a surface by nanoimprinting with a fine pitch mold. The moth eye mold has been prepared by electron beam (EB) lithography. However, EB lithography has the disadvantage of low throughput for preparing molds. It is so difficult to form a large moth eye mold that the technology to manufacture moth eye structures on large areas has not been realized.

10.4.2 Anodic Porous Alumina

Al anodizes in an acidic solution to form anodic porous alumina by self-organization, as illustrated in Fig. 10.3. Porous alumina is a typical naturally occurring ordered material. It has cells of a uniform size closely packed in a hexagonal manner. In a sample, every cell has the same size (within the range of 50–500 nm), the size controlled by the applied voltage, and has a pore one-third in size in the center. Generally, carbon is used as the cathode electrode and several acids, such as sulfuric acid and oxalic acid, are selected by applied voltage

[4]. Nanoimprinting processes based on the highly ordered anodic porous alumina have been extensively researched [5–7].

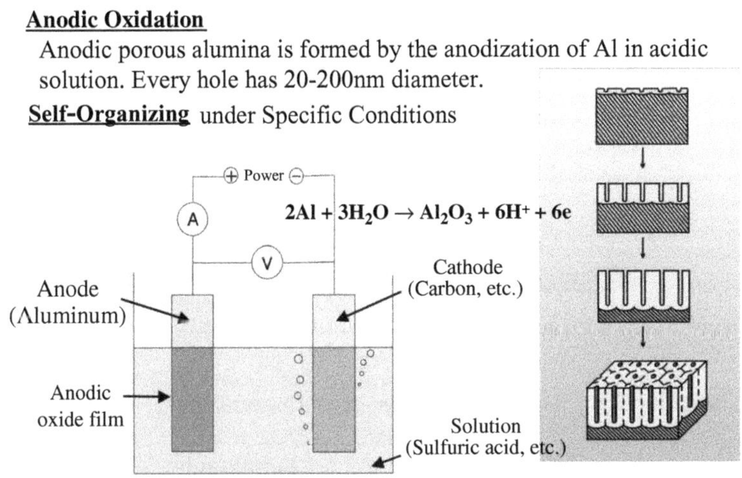

Figure 10.3 Formation of porous alumina.

10.4.3 Large Moth Eye Mold

Figure 10.4 illustrates the moth eye alumina mold production process. An oxalic acid water solution is used as an electrolyte. High-purity aluminum is anode-oxidized under a constant voltage and transforms into porous alumina. Next, porous alumina is etched with a phosphoric acid water solution to enlarge the pore diameters. By repeating these processes several times, a porous alumina mold with a taper shape is fabricated. Table 10.1 shows the characteristics of moth-eye-mold-producing processes. Lithography enables the production of structures involving small areas but is too expensive. And by lithography, we can't produce structures involving large areas or curved surfaces. On the other hand, porous alumina can be formed on large areas and on curved surfaces. Therefore, porous alumina can be realized as a big-sized roll mold required for the roll-to-roll process, with high productivity.

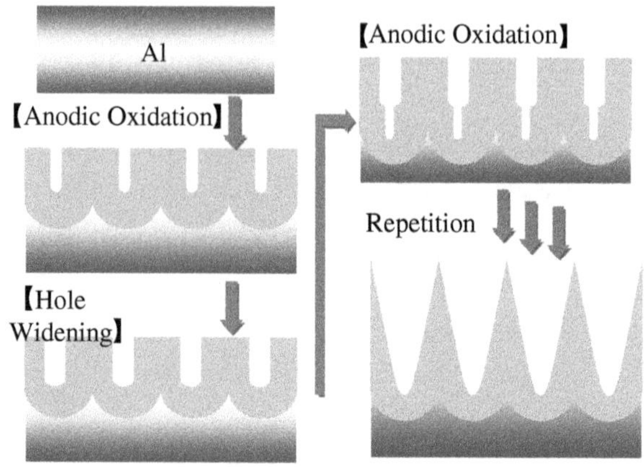

Figure 10.4 Moth eye alumina mold production process.

Table 10.1 Characteristics of moth eye–producing processes

	Anodic Porous Alumina	Lithography (Electron Beam Drawing, Interference exposure)
Structure Formation	Possible	Possible
Wide Area	Possible	Difficult
Roll Type Mold	Possible	Difficult
Production Cost	Low Cost	Expensive

10.4.4 Photonanoimprinting of the Moth Eye Structure

Figure 10.5 shows an image of a photonanoimprinting process with the moth eye mold [8]. First, the photopolymer is filled in the mold and the mold with the photopolymer is covered with a transparent substrate, for example, a polyethylene terephthalate (PET) film, a poly(methyl methacrylate) film, or a glass sheet. Next, ultraviolet (UV) light is irradiated over the covered transparent substrate; lastly, the moth eye sheet is detached from the mold.

Figures 10.6a and 10.6b are field-emission scanning electron microscopy (FE-SEM) images of the typical moth eye mold. Figure 10.6a is the cross-section image, and Fig. 10.6b is the surface image of the mold. Figure 10.6c shows the FE-SEM image of the typical moth eye surface on the PET film. It is confirmed that the shape of the taper porous alumina mold can be properly transferred as the moth eye surface.

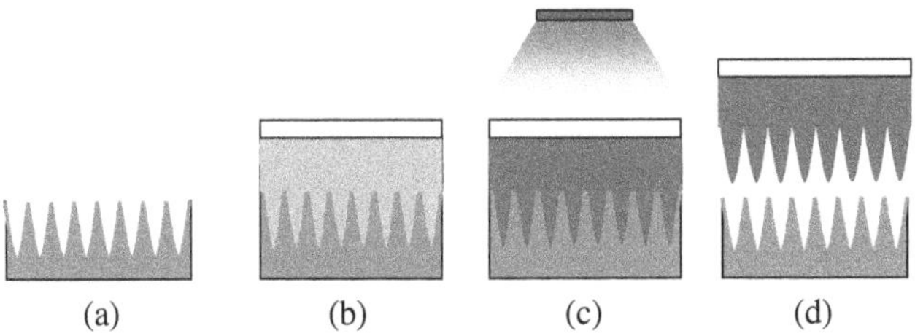

Figure 10.5 Photonanoimprinting process: (a) preparing anodic porous alumina, (b) filling with a photopolymer, (c) UV irradiation, and (d) detaching.

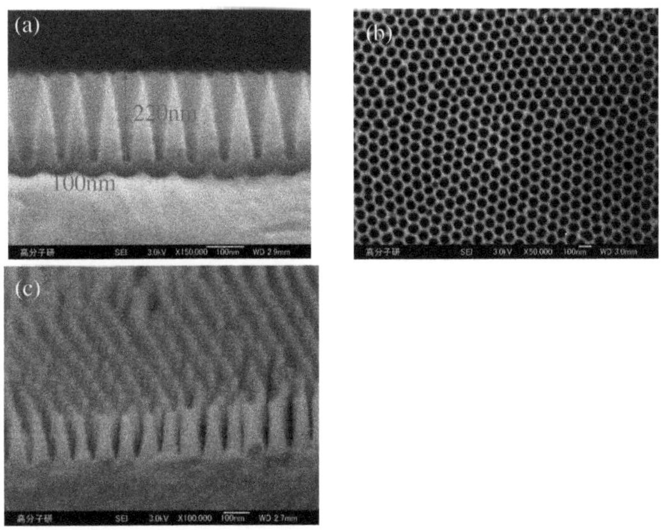

Figure 10.6 (a) Cross-sectional SEM image of a tapered alumina mold, (b) surface SEM image of the tapered porous alumina mold, and (c) cross-sectional SEM image of the moth eye surface.

10.4.5 Continuous Photonanoimprinting of the Moth Eye Structure

Figure 10.7 shows an image of continuous roll photoimprinting. The photopolymer is spread on the PET film. Then, the film is pressed on the roll mold and irradiated with UV light. The moth eye surface is made on the PET film continuously.

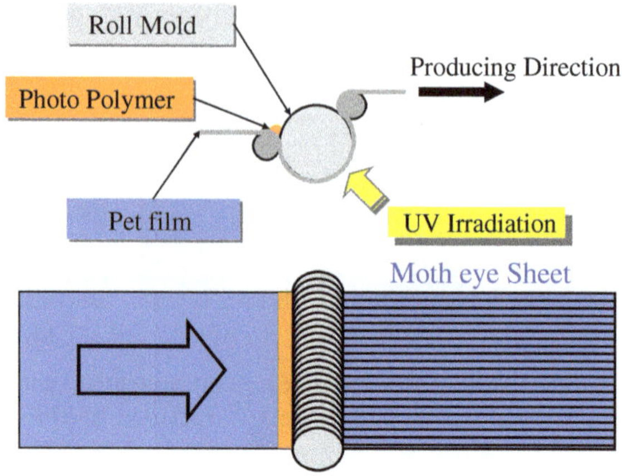

Figure 10.7 Image of continuous roll photonanoimprinting.

Figure 10.8a shows a seamless roll mold for the moth eye surface. Figure 10.8b is the photograph of the transferred moth eye film. The moth eye film is transparent, and the blue color comes from the protection film.

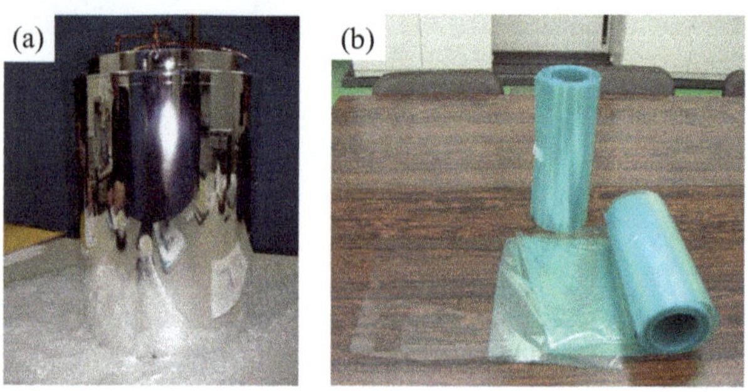

Figure 10.8 (a) Seamless roll mold for a moth eye surface and (b) transferred moth eye film.

10.5 Characteristics of the Moth Eye Antireflection Film

10.5.1 Antireflection

Figure 10.9 illustrates reflectance of a multilayer-type AR film and moth eye AR films at an incident angle of 5°. The horizontal axis is the wavelength of the incident light. Generally, multilayer-type AR films prevent reflection mainly at the wavelength that is the highest in visual sensitivity. On the other hand, moth eye AR films have lower reflection than the multilayer type and have low reflection throughout the visible wavelength.

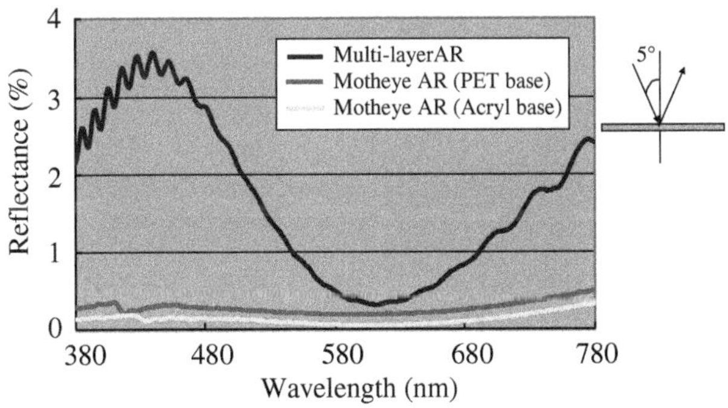

Figure 10.9 Reflectance of AR films for angle of incidence of 5°.

Figure 10.10 shows the incident angle dependence on reflectance. In this figure the horizontal axis is the angle of incidence. In the case of the multilayer-type AR film, reflectance becomes abruptly higher than 50°. For the moth eye AR film, the reflectance is lower than 1% even at 60°.

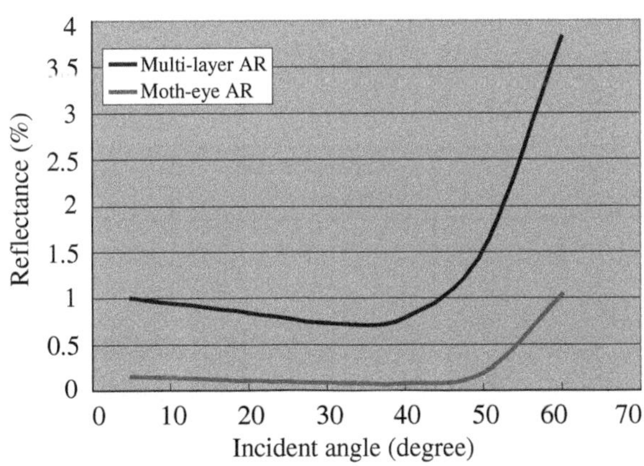

Figure 10.10 Incident angle dependence on reflectance at 550 nm.

Figure 10.11 shows the background reflections on the AR films. These images are the reflections of fluorescent lamps. The right is on the multilayer-type AR film, and the left is on the moth eye AR film. The right image has the apparent reflection of lamps. The left image has no reflection. It's confirmed that the moth eye AR film has a better performance.

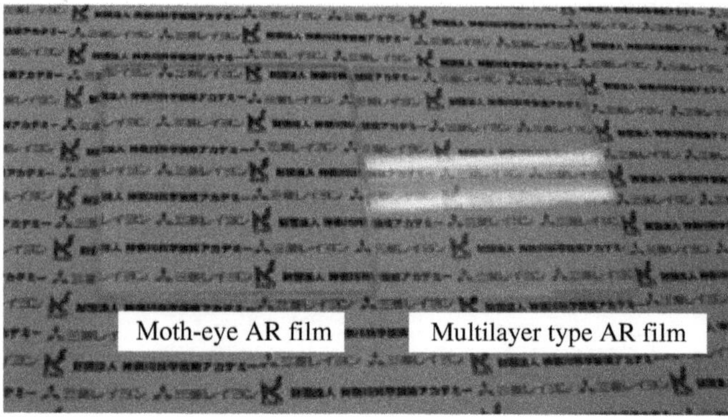

Figure 10.11 Background reflections on AR films.

10.5.2 Water Repellent

The moth eye surface made of a water repellent resin also has a superb water-repellent property. Figure 10.12 shows contact angles between water and several surfaces: glass, commercial water-repellent film, and moth eye AR film. For the moth eye AR film, the contact angle is over 140°. This phenomenon is known as the "Lotus effect."

	Glass	Commercial Water-Repellent Film	Moth-eye Film
Photographs	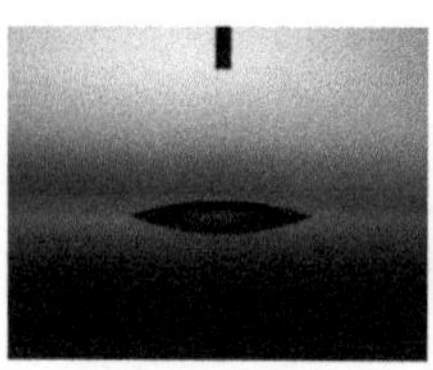		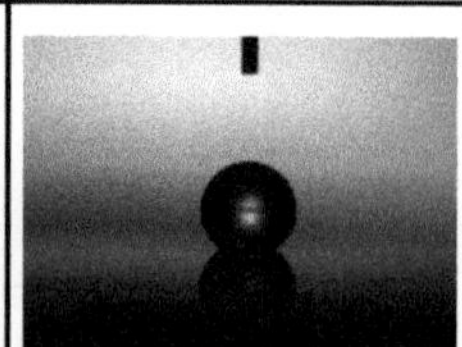
°	24	102	142

with 1 μL water

Figure 10.12 Contact angles with water of several surfaces.

10.6 Conclusion

The conclusions are as follows:

- Big-sized moth eye roll molds can be realized by anodization of Al rolls.
- Moth eye AR films are fabricated with moth eye roll molds continuously. The reflectance of the AR films is under 0.4%.
- A productive and low-cost production process has been realized by developing this process.

References

1. Dobrowolski, J. A., Poitras, D., Ma, P., Vakil, H. and Acree, M. (2002). Toward perfect antireflection coatings: numerical investigation, *Appl. Opt.*, **41**, pp. 3075–3083.
2. Clapham, P. B. and Hultley, M. C. (1973). Reduction of lens reflexion by the "moth eye" principle, *Nature*, **244**, pp. 281–282.
3. Ibn-Elhaj, M. and Schadt, M. (2001). Optical polymer thin films with isotropic and anisotropic nano-corrugated surface topologies, *Nature*, **410**, pp.796 –799.
4. Masuda, H. and Fukuda, K. (1995). Alumina nanoporous structures, *Science*, **268**, pp. 1466–1468.
5. Masuda, H., Yotsuya, M., Asano, M., Nishio, K., Nakao, M., Yokoo, A. and Tamamura, T. (2001). Self-repair of ordered pattern of nanometer dimensions based on self-compensation properties of anodic porous alumina, *Appl. Phys. Lett.*, **78**, pp. 826–828.
6. Yanagishita, T., Nishio, K. and Masuda, H. (2006). Nanoimprinting using Ni molds prepared from highly ordered anodic porous alumina templates, *Jpn. J. Appl. Phys.*, **45**(Part 2), pp. 29–32.
7. Yanagishita, T., Nishio, K. and Masuda, H. (2007). Polymer through-hole membrane fabricated by nanoimprinting using metal molds with high aspect ratio, *J. Vac. Sci. B*, **25**, p. L35.
8. Yanagishita, T., Yasui, K., Kondo, T., Kawamoto, Y., Nishio, K. and Masuda, H. (2007). Antireflection polymer surface using anodic porous alumina molds with tapered holes, *Chem. Lett.*, **36**, pp. 530–531.

Chapter 11

Transparent Superhydrophobic Film Created through Biomimetics of Lotus Leaf and Moth Eye Structures

Nobuyuki Itoh
Research and Development Center, Dai Nippon Printing Co., Ltd., 250-1 Wakashiba, Kashiwa, Chiba, 277-0871, Japan
Itou-N9@mail.dnp.co.jp

Biomimetics uses natural structures for human purposes. Two famous structures in nature, the lotus leaf and the moth eye, are extensively used in biomimetics. Both the lotus leaf and the moth eye have tiny convex structures, and they have antifouling and antireflection effects, respectively. The structure of the tiny nipple array was studied thoroughly. Furthermore, the wet states of the lotus leaf and the moth eye were also investigated on the basis of an optical experimental design result. A developed film that mimicked the moth eye showed the same superhydrophobic property as that of the lotus leaf when the surface was covered by free-energy fluorine materials. The results of the study were used to produce a highly transparent superhydrophobic film that repels water droplets

Industrial Biomimetics
Edited by Akihiro Miyauchi and Masatsugu Shimomura

ISBN 978-981-4800-07-5 (Hardcover), 978-0-429-05883-7 (eBook)
www.jennystanford.com

extremely well. Beyond well-known self-cleaning tiles, self-cleaning glass windows are becoming a reality.

11.1 Introduction

Biomimetics is expected to yield novel functions for all aspects of the industry. One of the most famous properties in nature is the superhydrophobic property observed on some plant leaves, such as the lotus flower and the sweet potato [1, 2]. These plants clean themselves by using rain or dew to maintain photosynthesis efficiency. The leaves of these plants consist of sharp conical pillars several hundreds of microns in size and wax on the leaf surface. The large surface area of the pillars enhances the wax effect, and hence, water droplets are repelled. This has been already applied to tiles and outer walls, rendering it difficult for them to become dirty.

Focusing on the fine pillar structure, the antireflective property of nanopillar arrays is another famous example of biomimetics observed in some species. Two examples are the surface of the compound eyes of moths and flies and the wing surface of cicadas and dragonflies [3, 4]. By preventing light from being reflected, these creatures effectively avoid being detected by predators.

There are large differences between the pillar sizes of the lotus leaf and the moth eye because their purposes are different, and these pillars are shown in Figs. 11.1 [1] and 11.2 [3]. If we mimic the surface structure of the lotus leaf for optical glazing or glass windows, these glass objects will not become transparent or hazy because the surface of the lotus leaf is rougher than the visible optical wavelength [5].

In this chapter, a transparent and superhydrophobic film is introduced by combining the super-water-repellent property of the lotus leaf and the nanopillar structure of the moth eye. This film can be used for optical glass windows.

First, the optical behavior of a nanopillar array is described, especially in terms of the reflection properties. Numerical simulations are performed for arrays with various diameters, pitches, heights, launch angles, and tip sharpness to determine the optimal antireflection condition (reflection ratio is less than 1%

for almost the entire visible wavelength with wide viewing angles). Subsequently, a thin layer of low surface free energy using fluorine compounds is deposited on a design-optimized antireflective nanopillar array. The relationship between the surface free energy and the wet state is discussed considering the fine surface structures.

Figure 11.1 (a) Lotus leaves, which exhibit extraordinary water repellency on their upper sides. (b) A scanning electron microscopy (SEM) image of the upper leaf side prepared using glycerol substitution shows a hierarchical surface structure consisting of papillae, wax clusters, and wax tubules. (c) Wax tubules on the upper leaf side. (d) Upper leaf side after critical-point (CP) drying. The wax tubules are dissolved; thus, the stomata are more visible. A tilt angle of 15°. (e) The leaf underside (CP dried) shows convex cells without stomata [1].

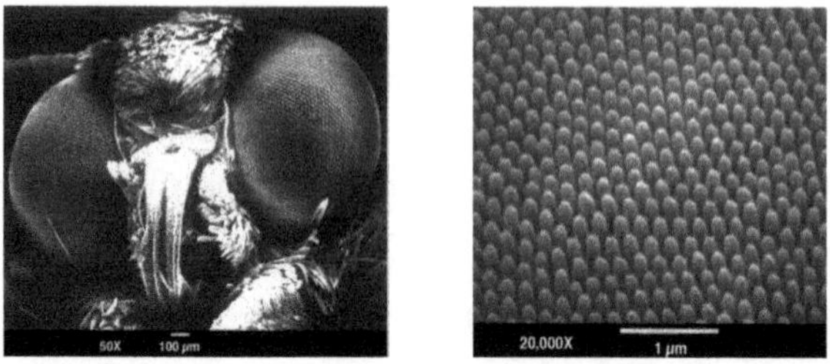

Figure 11.2 (Left) Eyes of a night-flying moth magnified 50 times. Note that each eye is composed of hundreds of lenslets arranged in a honeycomb pattern. (Right) Image showing surface structures observed to be covering the eyes of a moth, magnified 20,000 times. The texture consists of cone structures approximately 200 nm high packed into a hexagonal array with a spacing of 200 nm [3].

11.2 Optical Properties of a Transparent Superhydrophobic Film

Light refracts and reflects at the interface between air and a substance. Conical or tapered nanometer pillars of moth eye corneal nipple arrays were estimated to be multilayer structures with gradually varying refraction indexes [6, 7]. Consequently, there is no optical interface in these structures. The reflective behavior of these structures was simulated with various shapes and arrangements of nanopillar arrays in order to confirm the antireflection phenomenon. Optical properties were simulated using the diffraction mode of R-soft (Synopsys Inc.), which is used to design suboptical-wavelength periodic structures such as diffraction gratings. An electromagnetic wave analysis based on a rigorous coupled wave [8] was performed, and the reflection state was calculated.

The basic nanopillar array models used for the simulation were hexagonally and densely packed cones and circular truncated cones as illustrated in Figs. 11.3 and 11.5, respectively. These structures were derived from scanning electron microscope (SEM) images of the moth eye. Figure 11.3 shows a coordination model of the conical-shaped nanopillar structure of the moth eye used for the simulation. The refractive index is denoted as n and was assumed to be 1.58 and 1.5 for the base plate and resin-forming nanostructures, respectively. The thickness of the resin was assumed to be 1 μm. The height, bottom diameter, and pitch of the cones (not circular truncated cones) were 200 nm, 50 nm, and 100 nm, respectively. The incident angle, θ, of light is defined as shown in Fig. 11.3b.

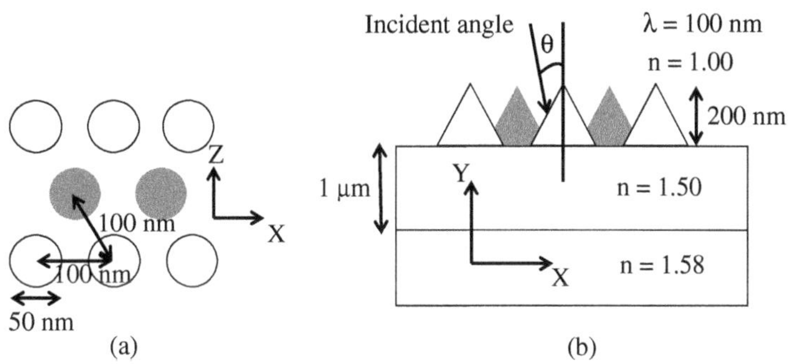

Figure 11.3 Coordination model of a nanopillar array for optical simulation. (a) Top view. (b) Cross section.

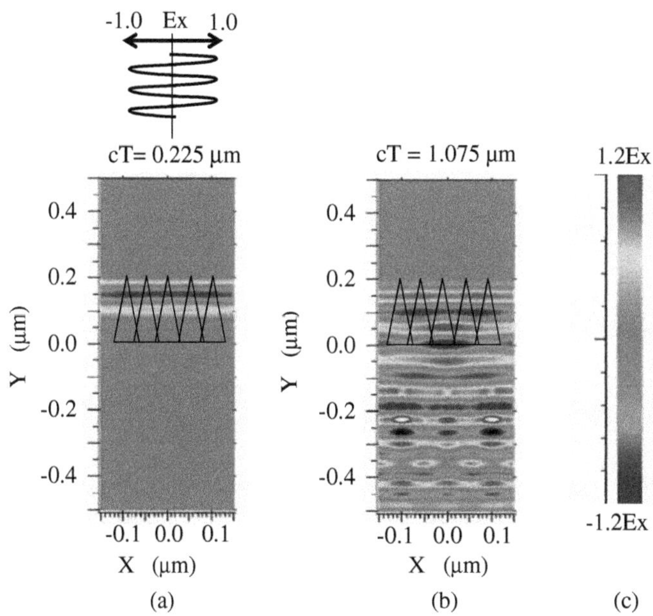

Figure 11.4 Simulated dynamic light propagation across a conical nanopillar array modeled after a moth eye. Distribution of the electric field amplitude. (a) The penetration depth is 0.225 μm. (b) The penetration depth is 1.075 μm. (c) Amplitude of a vibration light vector. It became slightly large via refraction and interference through the moth eye, compared to the incident light.

The simulated result of the dynamic light propagation is shown in Fig. 11.4. The *y* axis position of 0.2 μm indicates the interface of the cone tips. The light was expressed as the electric field vector E_x and propagated in the negative-*y* direction. The propagation position of light is expressed by the multiplication of celeritas *c* (m/s) and real time *T* (μs). An electromagnetic wave of length 100 nm was projected at an incident angle of 5°. It was observed that incident light penetrated the resin from the cone tips without any backward reflection.

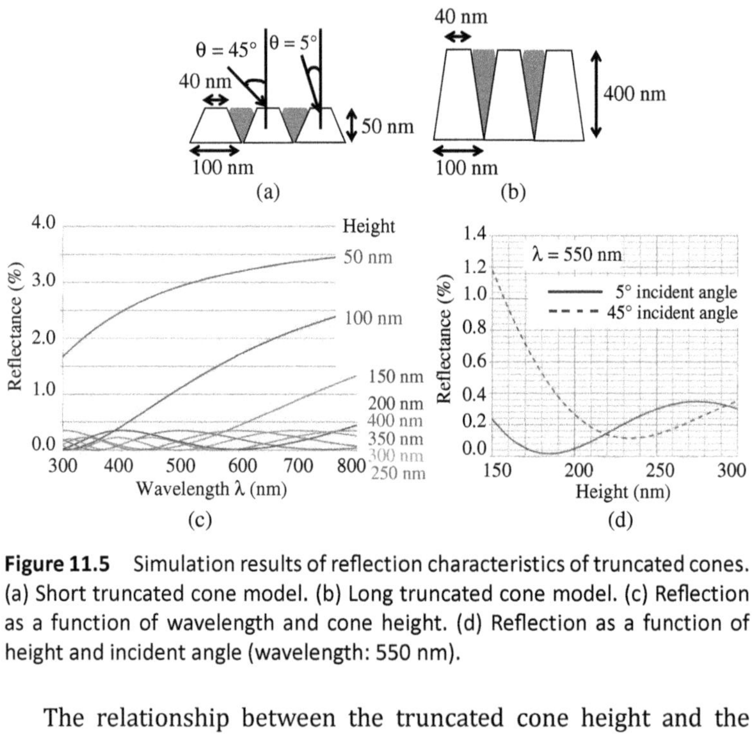

Figure 11.5 Simulation results of reflection characteristics of truncated cones. (a) Short truncated cone model. (b) Long truncated cone model. (c) Reflection as a function of wavelength and cone height. (d) Reflection as a function of height and incident angle (wavelength: 550 nm).

The relationship between the truncated cone height and the reflection property is shown in Fig. 11.5. The circular truncated cones closely aligned hexagonally at a diameter of 100 nm were modeled. The top and bottom diameters were 40 and 100 nm, respectively. The heights were in the range of 50 to 400 nm, as shown in Figs. 11.5a and 11.5b. Figure 11.5c shows the simulated reflection at the incident angle of 5° for various pillar heights. The reflection property strongly depended on the height of the circular truncated cone, and it was determined that a height of at least 200 nm is necessary for suppressing the reflection for the entire visible wavelength range. Figure 11.5d shows the incident angle dependence of the reflection property at a wavelength of 550 nm for varying pillar heights. It was observed that the lowest reflection was less than 0.1% at the angle of 5°. Furthermore, the dependence on the incident angle decreased when the pillar height was larger than approximately 200 nm.

The relationship between the steepness of the cone and the coloration in the reflection is shown in Fig. 11.6. The author defined the end fill factor as the ratio between the top and bottom diameters T/B, as shown in Fig. 11.6a, to express the steepness of the cones. The cone height was set to 200 nm, and the diameter of the top was changed from 0 to 55 nm, whereas that of the bottom was fixed at 100 nm, for which the end fill factor was 0 to 0.55, as shown in Figs. 11.6b and 11.6c. The incident angle was set to 5°. Figure 11.6d shows the simulated reflective spectrum for various end fill factors. The coloration in the reflection strongly depended on the end fill factor, as shown in Fig. 11.6d. The difference in reflectance among the wavelengths was small when the end fill factor was between 0.35 and 0.55, indicating neutral coloration.

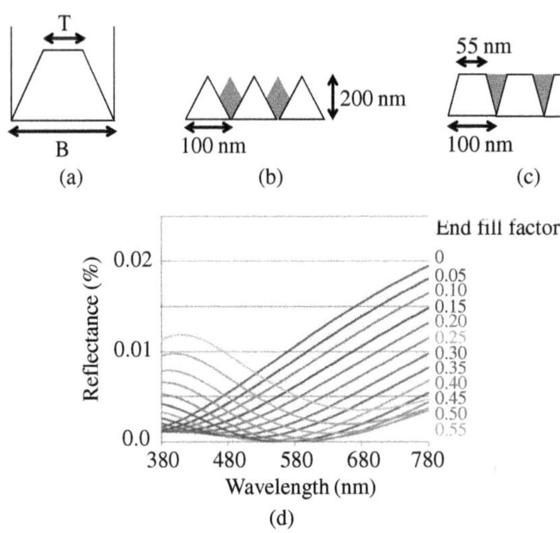

Figure 11.6 Cross section of a nanopillar array with various end fill factors for optical simulation and simulated results of reflection. (a) Definition of the end fill factor T/B. (b) Cross section of a cone with zero end fill factor. (c) Cross section of a truncated cone with an end fill factor of 0.55. (d) End fill factor and wavelength dependence of reflection.

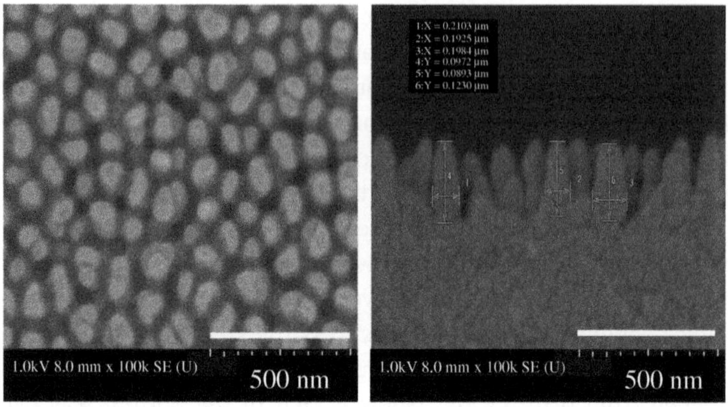

(a) (b)

Figure 11.7 SEM images of fabricated moth eye. (a) Top view. (b) Cross section.

To confirm the simulation result, a moth eye nanopillar array with a pitch of 100 nm, bottom diameter of 100 nm, end fill factor of 0.4, and height of 200 nm was fabricated. The launch angle was estimated to be approximately 5° in this case. A moth eye film was fabricated using a nanoimprinting method with a metal mold of millions of nanoholes. The surface of the metal was nanoetched by using an anode oxidization method, and the moth eye condition was controlled by setting the anode oxidization conditions. An ultraviolet (UV) curable urethane acrylate (UA) monomer was prepared and dropped on the metal mold, covered with a polyethylene terephthalate (PET) film, pressed to the mold, cured using UV, and released from the mold. An SEM image of the fabricated moth eye is shown in Fig. 11.7. The top view and cross section are shown in Figs. 11.7a and 11.7b, respectively. The reflection property of the moth eye is shown in Fig. 11.8 at an incident angle of 5°. The fabricated moth eye showed a very good antireflection property, the same as the simulation result over a wide visible wavelength range.

11.3 Study of Transparent Superhydrophobic Film Using a Moth Eye Nanostructure and Deposition of Fluorine Materials

The typical liquid states on a structured surface were investigated by Wenzel and Cassie and Baxter, as illustrated in Fig. 11.9 [9, 10].

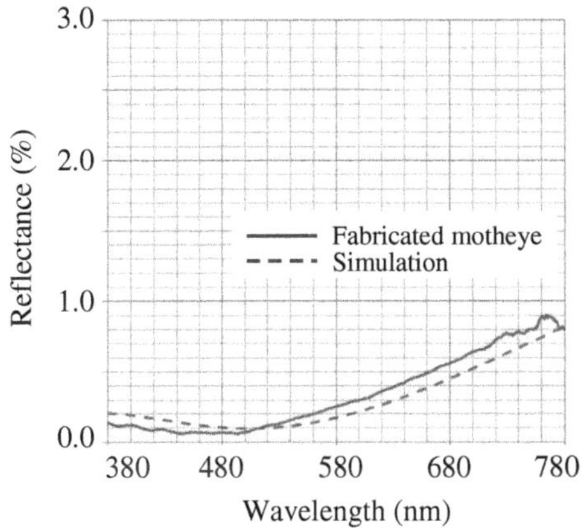

Figure 11.8 Reflection property of a fabricated moth eye compared with the simulation result at an incident angle of 5° over a wide visible wavelength range.

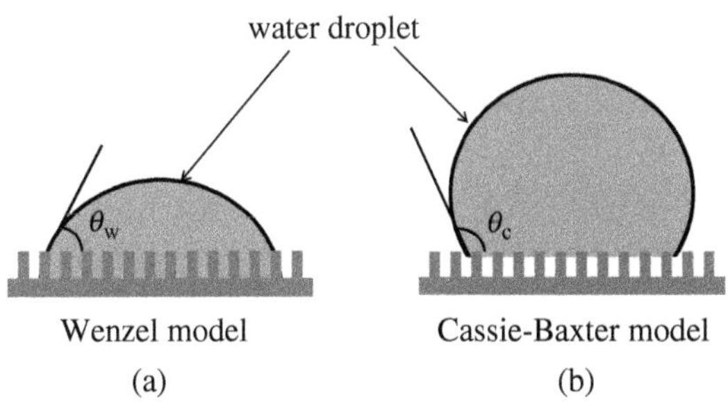

Figure 11.9 Illustration of wet states. (a) Hydrophilic Wenzel model. (b) Hydrophobic Cassie–Baxter model.

The wetting of a structured hydrophilic surface is often described by the Wenzel model, where the apparent contact angle is given by Eq. 11.1:

$$\cos\theta_W = \Gamma\cos\theta_0, \tag{11.1}$$

where Γ is the roughness factor, which is the ratio between the real and projected surface areas, and θ_0 is the intrinsic contact angle of the flat material. The wetting states with trapped air voids are described by the Cassie–Baxter model, where the apparent contact angle is given by Eq. 11.2:

$$\cos\theta_{CB} = \varphi(\cos\theta_0 + 1) - 1, \tag{11.2}$$

where φ is the ratio of the real liquid contact surface area to the apparent liquid contact surface area. Cassie reported a two-surface component system wetting in advance, described by Eq. 11.3:

$$\cos\theta_C = f_1\cos\theta_1 + f_2\cos\theta_2, \tag{11.3}$$

where f_1 and f_2 are fractions of materials 1 and 2 ($f_1 + f_2 = 1$) and θ_1 and θ_2 are the contact angles of intrinsic materials 1 and 2, respectively. Equation 11.2 was derived from Cassie's approach when the contact angle of water to air is 180°. The contact angle of water to air can result from a water droplet becoming round in a spacecraft when there is zero gravity. For a hydrophilic surface, that is, when the intrinsic contact angle $\theta_0 < 90°$, the surface structure is expected to increase the hydrophilicity of the surface according to the Wenzel model. For a hydrophobic surface, that is, when the intrinsic contact angle $\theta_0 > 90°$, the surface structure is expected to increase the hydrophobicity of the surface according to the Cassie–Baxter model. Superhydrophilicity is defined as the state at which the water contact angle is less than 10°, and superhydrophobicity is typically defined as the state at which the water contact angle is more than 150°.

The concentration of nanopillars of the fabricated moth eye was estimated to be to $45.4/\mu m^2$, as shown in Fig. 11.7a. The surface area ratio Γ of the Wenzel model was calculated to be 3 by considering the trapezoidal side. The ratio φ of the real liquid contact surface area to the apparent liquid contact surface area of the Cassie–Baxter model was 0.36 for the cylindrical shape, and hence, an accurate value of φ was obtained by calculating the product of the value of cylindrical shape and the end fill factors T/B, as shown in Fig. 11.6a.

Figure 11.10 shows the experimentally used fluorine compounds with CF_2 of different lengths, CF_3 with different terminal structures, etc. Chemical vapor deposition (CVD) was used to form a thin fluorine layer on the moth eye, maintaining the nanostructure.

Figure 11.11 shows the SEM images of a cross section of the moth eye on which (perfluorobutyl)ethylene was deposited. The CVD layer thickness was 26 nm, and the deionized water contact angle was 151° in Fig. 11.11a; whereas the CVD layer thickness was 98 nm, and the deionized water contact angle was 135° in Fig. 11.11b. A superhydrophobic state was obtained using a thin fluorine layer, one of less than 30 nm, which maintained the moth eye nanostructure.

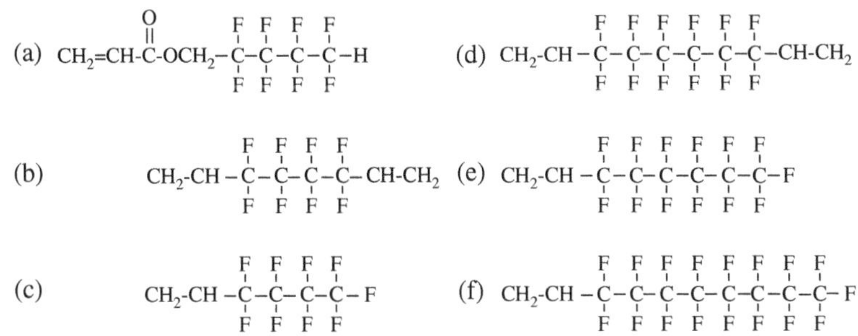

Figure 11.10 Experimentally used fluorine compounds with CF_2 and terminate structures of various lengths. (a) 1H,1H,5H-Octafluoropentyl acrylate from Osaka Organic Chemical Industry, Ltd. Its commercial name is Viscoat 8F and CAS no. is 376-84-1. (b) 1,4-Divinyl(perfluorotetra) from Tosoh Corporation. Its commercial name is C4-DV. (c) (Perfluorobutyl)ethylene from Daikin Industries, Ltd. Its commercial name is F-1420, and CAS no. is 19430-93-4. (d) 1,6-Divinyl(perfluorohexane) from Tosoh Corporation. Its commercial name is C6-DV and CAS No. is 1900-91-5. (e) (Perfluorohexyl)ethylene from Daikin Industries, Ltd. Its commercial name is F-1620, and CAS no. is 25291-17-2. (f) (Perfluorooctyl)ethylene from Tokyo Chemical Industry Co., Ltd. Its commercial name is H0846 and CAS no. is 21652-58-4.

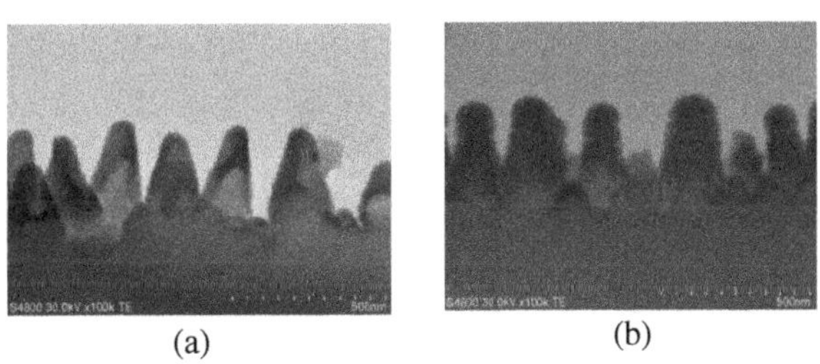

Figure 11.11 SEM images of the cross section of a moth eye on which a fluorine material, (perfluorobutyl)ethylene, was deposited. (a) A thin CVD layer was deposited. (b) A thick CVD layer was deposited.

Table 11.1 presents the contact angle and the surface free energy for both flat PET and the fabricated moth eye film covered with different fluorine compounds. By using the quartz crystal resonator film thickness meter in the CVD chamber, it was ensured that the fluorine layer thickness was less than 30 nm, measured by a contact-type film thickness meter using Si wafer references deposited at the same time for each sample. Number 1 indicates the results of the flat PET and flat UA, and Number 2 indicates the result of the UA

Table. 11.1 Contact angles and surface free energy for a flat PET film and a flat and fabricated moth eye UA film deposited with fluorine compounds

No.	Fluorine Compound	Number of		flat PET (no surface structure) Contact Angle (degree)		flat PET Surface Free Energy (mJ/m^2)			Urethane Acrylate (UA) Surface Structure	UA Contact Angle (degree)		UA Surface Free Energy (mJ/m^2)		
		CF2	CF3	DIW	n-HXD	γ	γd	γp		DIW	n-HXD	γ	γd	γp
1	—	0	0	67	2	40.9	27.5	13.4	No	56	5	47.9	27.5	20.4
2	—	0	0			—			moth eye	7	2		-	
3	1H,1H,5H-Octafluoropentyl Acrylate	4	0	70	21	38.1	25.7	12.4	moth eye	117	0		-	
4	1,4-Divinyl(Perfluorotetra)	4	0	94	38	24.8	22.0	2.8	moth eye	130	15		-	
5	(Perfluorobutyl)Ethylene	3	1	107	61	16.3	15.1	1.1	moth eye	156	91		-	
6	1,6-Divinyl(Perfluorohexane)	6	0	107	40	21.8	21.4	0.3	moth eye	129	5		-	
7	(Perfluorohexyl)Ethylene	5	1	110	67	14.2	13.3	0.9	moth eye	163	98		-	
8	(Perfluorooctyl)Ethylene	7	1	111	70	13.3	12.4	0.9	moth eye	155	85		-	

moth eye film. The six fluorine compounds in Fig. 11.10 were used to control the surface free energy of the outermost surface of the fabricated moth eye. Deionized water and *n*-hexadecane are denoted as DIW and *n*-HXD, respectively. The difference in surface free energy γ is derived from the contact angles of DIW and *n*-HXD solutions on the flat PET and flat UA by using a method developed by Owens and Wendt [11], where γd is the distribution component and γp is the polarity component of the surface free energy.

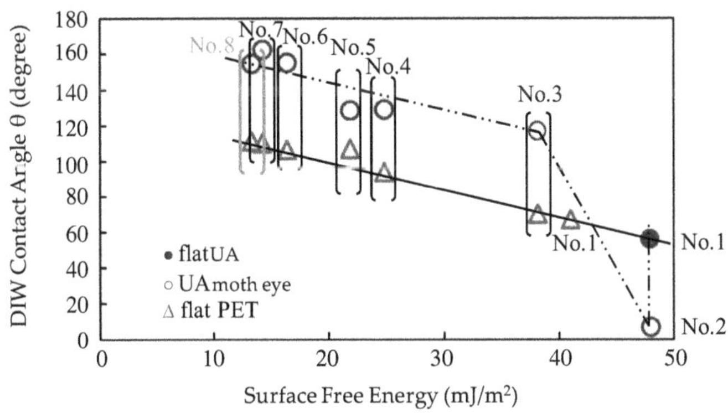

Figure 11.12 Relationships between surface free energy for the outermost layer and the water contact angle for different kinds of surfaces. The data numbers correspond to sample numbers in Table 11.1. The parentheses show the same fluorine compound pairs.

The relationship between the surface free energy obtained for flat surfaces and the contact angle for DIW of all the surfaces in Table 11.1 is shown in Fig. 11.12. There were three kinds of surfaces: flat PET, flat UA, and fabricated UA moth eye. Data numbers to which Table 11.1 corresponds are indicated in the data in Fig. 11.12. Pairs of flat PET and UA moth eye with the same fluorine compound are coupled using parentheses. The linearity between the surface free energy and the contact angle was evident for the flat surfaces, as shown by the solid line. The nanostructure of the moth eye causes the UA surface to be superhydrophilic once (• No. 1 to ○ No. 2) according to the Wenzel model. Subsequently, the contact angle increased sharply when the moth eye and the fluorine surface were combined, and it increased as the surface free energy decreased, as shown by the dotted line. The hydrophobic state was clearly enhanced by the moth eye according to the Cassie–Baxter model. The importance of the CF_3 terminal is presented in Fig. 11.13. Figure 11.12 was replotted for flat PET as the relationship between the number of fluorine molecules and the surface free energy γ to Fig. 11.13. The results that have a CF_3 terminal among them are in red. The surface free energy decreases with the number of fluorine molecules, and the existence of CF_3 is more effective than $(CF_2)_n$.

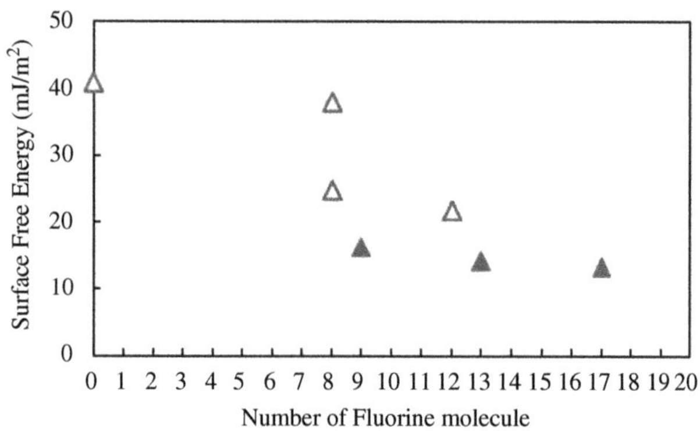

Figure 11.13 Relationship between the number of *F* and surface free energy for flat PET. The results that have a CF_3 terminal and only CF_2 among them are in filled and open triangles, respectively.

Figure 11.14 shows the relationship between the surface area ratio Γ and the apparent contact angle θ_w in the Wenzel model calculated for various intrinsic contact angles of θ_0, followed by Eq. 11.1. The Wenzel model suggests that the fabricated surface easily changed to a superhydrophilic state when the intrinsic contact angle θ_0 was low and the surface area ratio Γ was large. The intrinsic contact angle θ_0 of the UA for DIW was 56° at a Γ of 1, and the contact angle of the UA moth eye without fluorine, θ_w, was 7° at a Γ of 3. These are plotted in Fig. 11.14 (• and ○). It was observed that the condition of the UA moth eye was sufficient for the superhydrophilic state.

The wet state property of the superhydrophobic moth eye film is shown in Fig. 11.15. The fluorine compound 3-perfluorohexyl-1,2-epoxypropane was deposited on the film. A colored aqueous solution was prepared using 0.5 wt% methylene blue. High transparency and a superhydrophobic state could be easily observed.

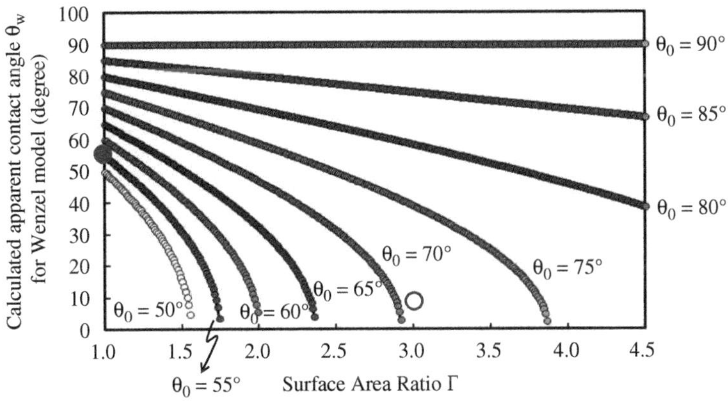

Figure 11.14 Simulated relationship between surface area ratio Γ and apparent contact angle θ_W of the Wenzel model calculated for various intrinsic contact angles θ_0. The intrinsic conditions of flat UA (•) and UA moth eye condition (o) are plotted.

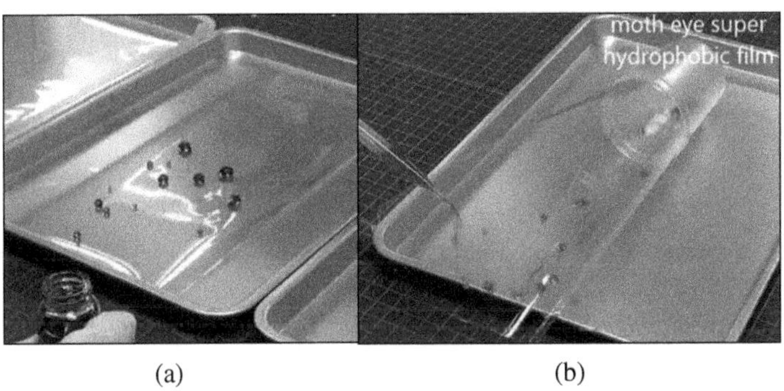

Figure 11.15 Transparent superhydrophobic film. (a) Water droplets colored with methylene blue balled up like beads. (b) Water droplets slid down as though on a roller coaster when the film was rounded.

11.4 Summary

Biomimetic nanostructure surfaces were studied from the aspects of optics and wettability. A moth eye design was optically optimized. The conical shape of the moth eye surface pillars, which is defined by the end fill factor, is important in addition to the usual nanostructure parameters, such as diameter, pitch, and height. It was determined that the tiny conical shape of the moth eye is also necessary for enhancing water repellency. In summary, a highly transparent, superhydrophobic film was developed by using nanostructures of the moth eye covered with a thin layer of long-chain fluorine compounds by applying the superhydrophobic effect of the lotus leaf. The film will be useful for creating self-cleaning windows. The author envisages that biomimetics might lead to novel ideas and value beyond simple replication of nature.

Acknowledgments

The author wishes to thank Shogo Kubota, Osamu Iwata, Taro Morimoto, and Kaori Yamashita of Dai Nippon Printing Co., Ltd., for their help both in the experiment and the discussion.

References

1. Ensikat, H. J., Kuru, P. D., Neinhuis, C. and Barthlott, W. (2011). Superhydrophobicity in perfection: the outstanding properties of the lotus leaf, *Beilstein J. Nanotechnol.*, **2**, pp. 152–161.
2. Koch, K. and Barthlott, W. (2009). Superhydrophobic and superhydrophilic plant surfaces: an inspiration for biomimetic materials, *Philos. Trans. R. Soc. London, Ser. A*, **367**, pp. 1487–1509.
3. MacLeod, B. D. and Hobbs, D. S. (2004). Low cost anti-reflection technology for automobile displays, *Proc. SID Vehicle Display Conference October 2004*, **2.6**, pp. 1–6.
4. Han, Z. W., Wang. Z., Feng, X. M., Li, B., Mu, Z. Z., Zhang, J. Q., Niu, S. C. and Ren, L. Q. (2016). Antireflective surface inspired from biology: a review, *Biosurf. Biotribol.*, **2**, pp. 137–150.
5. Sun, M., Luo, C., Xu, L., Ji, H., Ouyang, Q., Yu, D. and Chen, Y. (2005). Artificial lotus leaf by nanocasting, *Langmuir*, **21**, pp. 8978–8981.
6. Clapham, P. B. and Hutley, M. C. (1973). Reduction of lens reflection by the 'moth eye' principle, *Nature*, **244**, pp. 281–282.
7. Mukherjee, A. (2010). *Biomimetics Leaning from Nature*, eds. Lohmueller, T., Brunner, R. and Spatz, J. P., Chapter 22; Improved properties of optical surfaces by following the example of the 'moth eye' (In Tech Europe, Croatia), pp. 451–466.
8. Bodena, S. A. and Bagnall, D. B. (2008). Tunable reflection minima of nanostructured antireflective surfaces, *Appl. Phys. Lett.*, **93**, p. 133108.

9. Wenzel, R. N. (1936). Resistance of solid surfaces to wetting by water, *Ind. Eng. Chem.*, **28**, pp. 988–994.
10. Cassie, A. B. D. and Baxter, S. (1944). Wettability of porous surfaces, *Trans. Faraday Soc.*, **40**, pp. 546–551.
11. Owens, D. K. and Wendt, R. C. (1969). Estimation of the surface free energy of polymers, *J. Appl. Polym. Sci.*, **13**, pp. 1741–1747.

Chapter 12

Adhesion under Wet Conditions Inspired by Marine Sessile Organisms

Motoyasu Kobayashi
Department of Applied Chemistry, School of Advanced Engineering, Kogakuin University, 2665-1 Nakano-cho, Hachioji, Tokyo 192-0015, Japan
motokoba@cc.kogakuin.ac.jp

Artificial adhesives or coating materials inspired by marine sessile organisms have attracted much attention due to their useful application under wet conditions, such as for surgical tissue adhesives and environmentally friendly adhesives. Mussels are able to anchor themselves to rocks or substrates, even under seawater, by secreting adhesive proteins containing 3,4-dihydroxy-*L*-phenylalanine (DOPA) from their foot threads to form hydrogel-like adhesive pads through hydrogen bonding and chelating between the dihydroxy phenyl group and the substrate. The dihydroxy phenyl group in DOPA can also be found in sea squirts and urushi, producing a nice coating material through oxidative cross-linking reaction. Barnacles latch onto the surface of various materials through multifunctional coupling, including both polar and hydrophobic interaction, with

Industrial Biomimetics
Edited by Akihiro Miyauchi and Masatsugu Shimomura

ISBN 978-981-4800-07-5 (Hardcover), 978-0-429-05883-7 (eBook)
www.jennystanford.com

the substrate surface and hydrogel formation based on the self-assembly of peptide groups. This chapter describes recent works on several types of synthetic adhesive polymers and surface coating materials with DOPA moiety or catechol derivatives inspired by the chemical structure of adhesive proteins and hydrogels of marine sessile organisms, such as mussels, sea squirts, and barnacles.

12.1 Introduction

Various types of artificial adhesives and glues are commercially available today, but most of them do not work well under water or on wet surfaces. Dry, clean surfaces are generally necessary for joining materials by adhesive or glue, because even a small amount of liquid or contamination can disrupt the interaction between the substrate and the adhesive at the molecular level, resulting in low adhesion strength. On the other hand, marine organisms, such as mussels and barnacles, are able to adhere to a rock surface under seawater by natural means. Therefore, understanding the adhesion mechanism of sessile marine organisms is expected to be useful for creating new artificial adhesives that work under wet conditions, as in the case of surgical tissue adhesives, as well as nontoxic and environmentally friendly adhesives. Figure 12.1 shows typical living organisms that have attracted attention due to their useful adhesive mechanism and chemical components.

For example, mussels anchor themselves on a rock surface by secreting adhesive proteins containing 3,4-dihydroxy-*L*-phenylalanine (DOPA) [1] from their foot threads. The key chemistry for adhesion is the catechol dihydroxy phenyl group in DOPA. One of its features is the hydrogen bonding and chelating of two hydroxyl groups of catechol units with a metal oxide in the substrate to promote adhesion. The other feature is the oxidative cross-linking reaction under basic aqueous conditions ($pH > 8$), forming a hydrogel to improve the mechanical strength of the adhesion layer. Catechol derivatives, such as dopamine, norepinephrine [2], pyrogallol, tannin [3], and urushiol [4, 5], as shown in Fig. 12.1, also induce an oxidative cross-linking reaction to give adhesive properties similar to or better than catechol. Therefore, living things that have the catechol group, such as mussels, sea squirts, and urushi, have also attracted considerable attention for the design of new ecofriendly adhesives. Barnacles do not use the catechol chemistry in the attachment process. A barnacle anchors itself with a cement protein consisting of a multiprotein complex and a peptide-based hydrogel that self-assemble in a noncovalent fashion, a concept that has already been applied to the design of adhesive polymers. The following sections introduce some recent developments in adhesives and coating materials inspired by bioadhesion concepts originating from natural life forms.

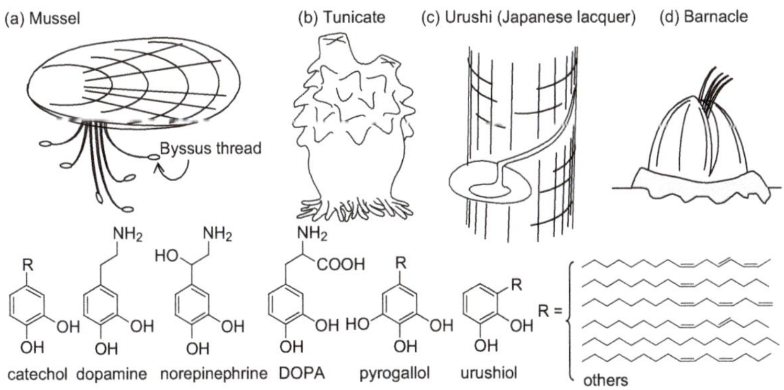

Figure 12.1 Source living organisms (upper) and related chemical components (lower) that have contributed to recent works on bioinspired adhesives and coating materials; (a) mussel, (b) tunicate, (c) urushi, and (d) barnacle and the chemical structures of catechol, dopamine, norepinephrine, DOPA, pyrogallol, and urushiol.

12.2 Mussel-Inspired Adhesives

Mussels anchor themselves on a rock surface by using byssus threads made of fibrous proteins [6]. Figure 12.2 shows an illustration of the threads and the chemical structure of the related adhesive proteins. Several types of mussel foot proteins (mfps) are secreted from the foot threads to form an adhesive plaque for binding with the substrate surface. Six types of proteins (mfp-1 to mfp-6) are connected to each other in the plaque [7–10]. Each protein contains lysine and the "nonstandard" amino acid DOPA, hydroxyl, and amino groups. The catechol group in DOPA plays two roles: adhesion with the substrate through hydrogen bonding or chelation [11, 12] and oxidative cross-linking reaction forming a hydrogel [13], as shown in Fig. 12.3. The catechol hydroxyl groups afford hydrogen bonding for the mfps to interact with metal oxides, such as SiO_2 and TiO_2 [14, 15], and also coordination with metal ions, in particular Fe^{3+} [16]. In a basic aqueous environment (pH > 8), a catechol group loses two electrons and two protons to become a highly reactive catechol quinone, which subsequently reacts to form cross-linked products via several pathways. One is a Schiff base formation reaction or a Michael-type addition of amines to the quinone, which resembles melanin formation. The other is a radical coupling of catechol quinone. The resulting cross-linked structures contribute sufficient mechanical adhesion strength for the adhesive plaque even under water.

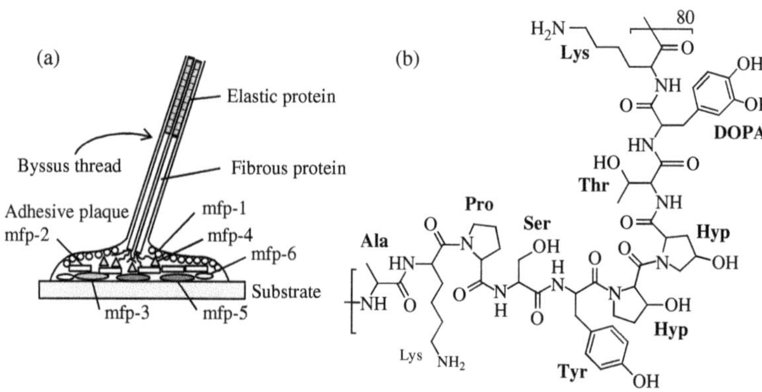

Figure 12.2 (a) Adhesion proteins (mfp-1 to mfp-6) in the adhesive plaque of mussel and (b) typical chemical structure of mfp-1, containing DOPA, hydroxyl, and amino groups.

Toward the useful application of mussel adhesive proteins (MAPs), various types of polypeptides have also been synthesized to demonstrate underwater adhesion. Yamamoto et al. chemically synthesized a model polypeptide that had a natural mfp-1 sequence to demonstrate the adhesion of two metal pieces using an aqueous solution of the polypeptide [17]. Deming and coworkers carried out ring-opening polymerization of *N*-carboxyanhydride monomers to produce an MAP mimetic adhesive copolymer composed of lysine and DOPA, as shown in Fig. 12.4a [18]. They investigated the effect of components of amino acid and an oxidizing agent, such as oxygen, tyrosinase, Fe^{3+}, H_2O_2, or periodate (IO_4^-), on the adhesion strength of the resulting cross-linked copolymer. Cha et al. obtained large quantities of recombinant hybrid proteins comprising mfp-1 by biosynthesis using *Escherichia coli* [19]. The bulk scale adhesion of the recombinant mussel adhesive hybrid was shown to achieve shear adhesive strength of 0.86 MPa when periodate was used as the cross-linking reagent.

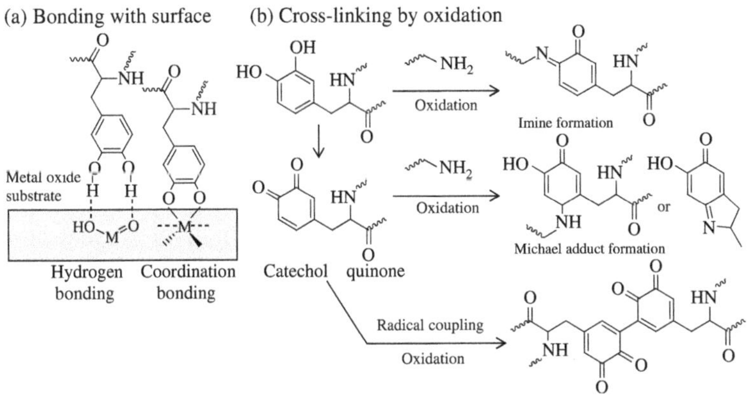

Figure 12.3 Adhesion mechanism of DOPA-containing mussel adhesive proteins by (a) bonding with the substrate surface via hydrogen bonding and coordination to metal oxide and (b) cross-linked hydrogel formation by oxidation of DOPA units, including imine formation, Michael adduct formation, and radical coupling reaction.

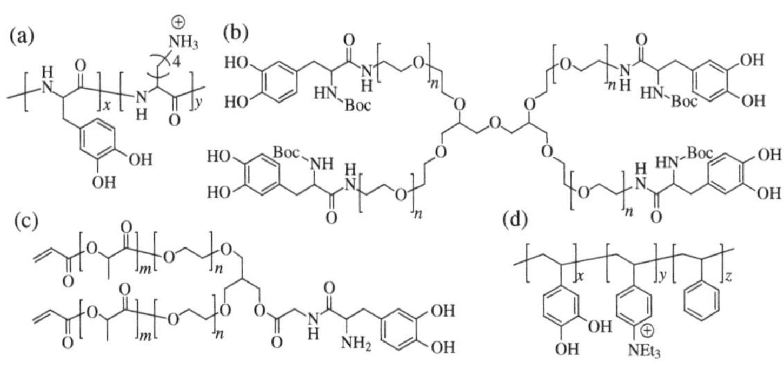

Figure 12.4 Chemical structure of MAP mimetic polymers used for adhesive and sealant; (a) poly(lysine) bearing dopamine units, (b) branched poly(ethylene oxide) (PEO), (c) diacrylate-type macromonomer bearing a catechol unit, and (d) MAP mimetic polystyrene copolymer.

MAP mimetic synthetic polymers have also been investigated. Lee et al. synthesized branched poly(ethylene oxide) (PEO) containing four DOPA end groups [20], as shown in Fig. 12.4b, for which oxidative polymerization resulted in cross-linked hydrogel formation. Depending on the presence or absence of an *N*-protective group (*t*-butoxy carbonyl: Boc) at the terminal amine, radical coupling or melanin formation mainly proceeded to form a cross-link. Messersmith et al. carried out photopolymerization of diacrylate-type macromonomers having a catechol group connected through PEO segments (Fig. 12.4c) in aqueous conditions to give a hydrogel [21]. The work of adhesion of the hydrogel and the contacting titanium substrate in water was measured by a contact mechanics method. Wilker et al. synthesized catechol-containing polystyrene (Fig. 12.4d), which revealed better lap shear adhesive strength of aluminum substrates submerged in artificial seawater compared to popular commercial glues consisting of poly(vinyl acetate) or cyanoacrylate [22, 23].

To improve the user friendliness of MAP mimetic adhesives, control of gelation is very important. Although the oxidation reaction of catechol is controllable by the pH or oxidant concentration, additional molecular design is necessary to ensure gelation at the desired time and position, quick reaction, and stable storage of the adhesive polymers before use. Photopolymerization of the macromonomer as mentioned in Fig. 12.4c achieved rapid curing within one minute to form an adhesive hydrogel owing to the catechol units and divinyl groups. Nishida et al. prepared MAP mimetic hydrophilic copolymer bearing caged catechol units using an *ortho*-nitrobenzyl protecting group [24], which can be readily deprotected by light irradiation. As shown in Fig. 12.5, a pH 8 phosphate buffer solution (PBS) of caged MAP mimetic copolymer was sandwiched between two glass plates, and then light was irradiated in a humid air atmosphere. The light deprotection of the caged compound gave a native catechol group and triggered an oxidative cross-linking reaction to afford the adhesive gel [25], resulting in adhesion of the glass plates.

Thermosensitive biocompatible hydrogel tissue adhesives have also been prepared by combining the catechol-conjugated chitosan and thiol-terminated pluronic copolymers [26, 27]. The two copolymers were dissolved separately in a pH 7.4 PBS solution, premixed for 1 h at 4°C and injected to use at 37°C, as shown in Fig. 12.6. The cross-linking reaction between the chitosan and pluronic copolymers was accelerated at 37°C by the Michael-type addition reaction of catechol quinone with the amine and thiol groups. Biomedical application of these copolymers has already been investigated for in situ sealing of punctured tissues to prevent bleeding [28].

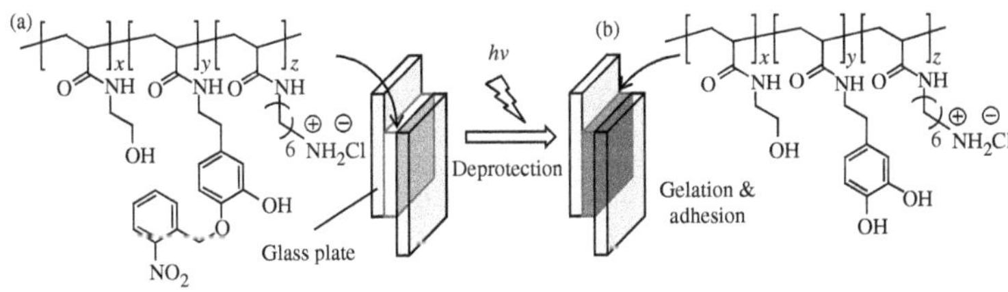

Figure 12.5 Photosensitive adhesion using an MAP mimetic copolymer.

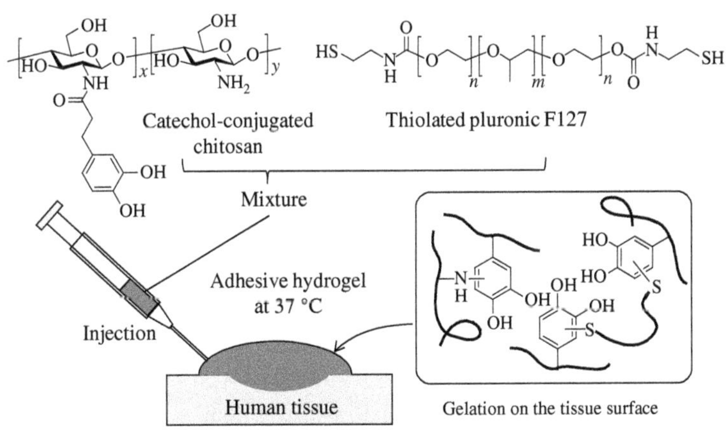

Figure 12.6 Tissue adhesive, thermosensitive, and in situ cross-linkable catechol-conjugated chitosan/thiolated pluronic F127 hydrogel.

Control of the gelation pathway of catechol-containing polymers has also been investigated. In most cases, catechol forms a covalent-type cross-linked gel via oxidative coupling in a basic environment. On the other hand, catechol-metal ion complexes are often observed in nature, such as marine mussel adhesive fibers [29, 30] and siderophores of enteric bacteria [31]. The chelation of the catechol dihydroxy group to metal ions also works as a cross-linking junction by bis-catecholate or tris-catecholate complex formation in basic media [32]. Using pH-sensitive coordination, a reversible cross-linking system has been reported. As shown in Fig. 12.7, a polystyrene copolymer bearing catechol side chains was oxidized in sodium methoxide (CH_3ONa) tetrahydrofuran (THF) solution to produce an irreversible gel via oxidative coupling reaction. With the presence of Fe^{3+} in the CH_3ONa/THF solution, cross-linking by coordinative coupling was dominant because the coordination was faster than the oxidation. Therefore, the resulting gel of the catechol-containing copolymer made it possible to reproduce the original copolymer by the addition of acetic acid [33]. The reversible cross-linking via metal coordination of catechol has been applied to self-healing polymer systems [34, 35].

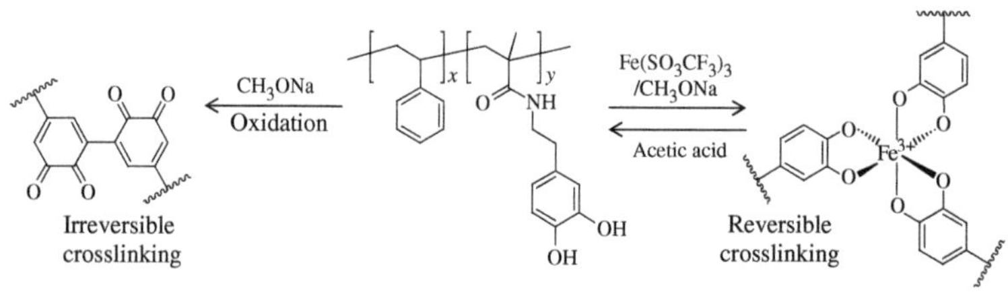

Figure 12.7 Dual gelation of dopamine-containing polymer (center) in THF containing CH_3ONa by oxidation based on irreversible cross-linking (left) and reversible coordination with Fe^{3+} ion (right), which reproduce the original copolymer by acetic acid.

Interestingly, when the substrate was immersed in the dopamine aqueous solution, dopamine was readily adsorbed on the substrate surface to form a unique polydopamine thin film by the successive oxidation and cross-linking reaction, as shown in Fig. 12.8a. The polydopamine film is formed on the surface of various substrates, such as gold, stainless steel, titanium, glass, silicon, poly(dimethyl siloxane) (PDMS), and poly(tetrafluoroethylene) [36]. Since the catechol group works as a useful anchoring function, various catechol-containing coating materials have been designed [37]. For example, catechol end–capped PEO, shown in Fig. 12.8b, was synthesized by condensation reaction of amino-terminated PEO and DOPA and was coated on gold and titanium oxide to give an antifouling surface that prevents the attachment of living cells

[38]. Messersmith et al. synthesized a hydrophilic polyacrylate copolymer bearing a catechol group (Fig. 12.8c) that could become immobilized on the top surface of microfabricated pillars made of PDMS [39], which exhibited repeatable adhesion and smooth release both in air and water, like a gecko's finger pads. Lee et al. prepared catechol-containing heparin (Fig. 12.8d); the surface of an artificial blood vessel made of polyurethane was dip-coated with this to improve the blood compatibility and prevent blood coagulation and platelet adhesion due to heparin anchored on the surface by catechol [40]. In addition, the polydopamine was used as a precursor film for polymer brush preparation. As shown in Fig. 12.8e, a 2-bromoisobutyloyl group was immobilized on the substrate surface by using catechol film formation [41]. The alkyl bromide can initiate radical polymerization of vinyl monomers in the presence of a copper catalyst to produce polymers, referred to as atom transfer radical polymerization (ATRP). Surface-initiated ATRP (SI-ATRP) of oligo-ethylene glycol methacrylate produces a densely grafted brush-like polymer to give a hydrophilic and antifouling surface. Recently, a polydopamine film coating using a spray system was developed to significantly accelerate the speed of cross-linking and immobilization [42].

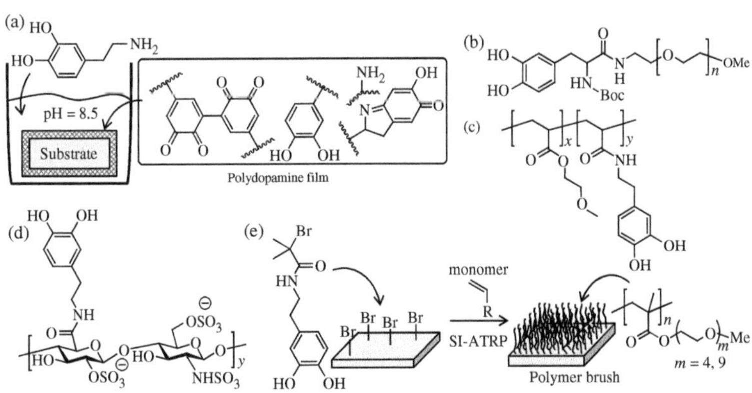

Figure 12.8 Chemical structure of MAP mimetic polymers used for surface coating; (a) polydopamine film formed by dopamine oxidation, (b) end-functionalized PEO attached to catechol, (c) catechol-containing hydrophilic copolymer, (d) catechol-containing heparin, (e) fabrication of polymer brush by SI-ATRP from a catechol-functionalized surface initiator.

12.3 Tunicate-Inspired Adhesives

"Sea squirts" is the common name for ascidians, which are sessile marine organisms that sometimes release jets of water through their siphons when disturbed. Ascidians belong to tunicates, having a tough body composed of "tunicin," a highly crystalline cellulose microfibril [43, 44] composite of carbohydrate and acid mucopolysaccharide, including 3,4,5-trihydroxyphenylalanine (TOPA), and pyrogallol amino acid. The pyrogallol units are considered to form covalent cross-links under a seawater environment (pH 8.2) and to enhance the mechanical strength of tunicin. In addition, one of their unique features is vanadium storage in a biological system as a complex of TOPA in their blood cells [45–48]—metal coordination that must be useful as a cross-linking junction. Furthermore, tunicates also contribute to tunic wound repair on ascidians. Blood cells aggregate at the wounded area of the tunic to help the blood coagulate and heal the injury. Thus, various applications of pyrogallol as well as TOPA are expected.

Inspired by the pyrogallol chemistry and the nanofiber structure of the tunicate, Oh et al. synthesized tunicate mimetic hydrogel adhesives using chitin nanofibers modified with gallic acid, followed by oxidation or metal complex formation, as shown in Fig. 12.9 [49].

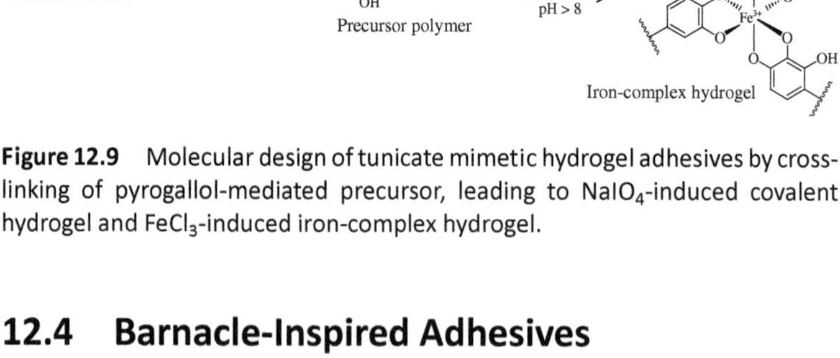

Figure 12.9 Molecular design of tunicate mimetic hydrogel adhesives by cross-linking of pyrogallol-mediated precursor, leading to $NaIO_4$-induced covalent hydrogel and $FeCl_3$-induced iron-complex hydrogel.

12.4 Barnacle-Inspired Adhesives

Figure 12.10a shows barnacles attached to material surfaces in seawater through the formation of a proteinaceous adhesive cement

[50, 51]. In fact, the adhesive materials and adhesion mechanism have not been fully understood due to the extremely low solubility of the adhesive cement. The cement is a multiprotein complex consisting of six amino acid–based proteins, a charged amino acid–rich protein, remarkably hydrophobic proteins, and an enzyme. The features of barnacle adhesives reported until now are that (i) cross-linking joints are not covalent bonds [52, 53] and (ii) proteins localize at the interface of the adhesive and substrate containing hydrophilic and hydrophobic amino acid residues [54, 55] that interact with various substrates. As shown in Fig. 12.10b, amyloid-like aggregates of self-assembled proteins are expected to produce an insoluble and tough cross-linking network.

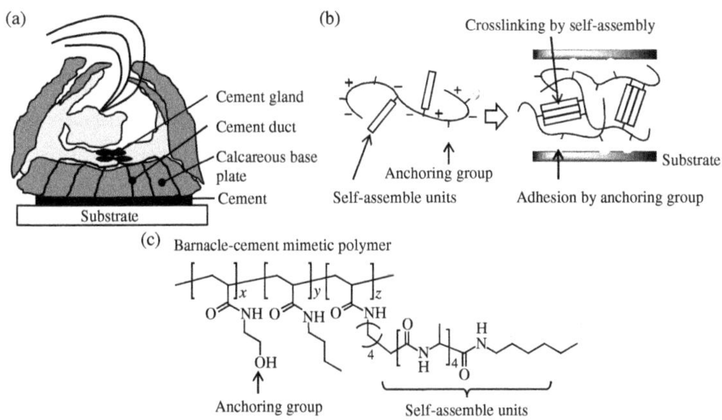

Figure 12.10 Schematic view of (a) barnacle attachment using the cement layer, (b) expected adhesion mechanism for the barnacle cement, and (c) chemical structure of the barnacle cement mimetic polymer.

Inspired by a barnacle adhesive protein mimetic, Nishida et al. designed and synthesized a new poly(acrylamide) for underwater adhesion consisting of a self-assembly motif for polymer cross-linking and surface anchoring moieties, as shown in Fig. 12.9c [56]. Tetraalanine units with a hydrophobic tail of hexyl groups were introduced in the polymer to induce a self-assembly behavior in solution. Hydroxyethyl and butyl groups at the side chains are expected to have both polar and van der Waals interactions with the substrate surface. An aqueous solution of the copolymer in the presence of condensation reagents and curing reagent (hexylamine) was sandwiched between two substrates to form a hydrogel after 4 h. Adhesion of stainless steel plates and PMMA plates was accomplished through the gelation of the aqueous polymer solution.

12.5 Summary

This chapter described underwater adhesives or coating materials inspired by marine sessile organisms, such as mussels, tunicates, and barnacles, in particular, MAP mimetic polymers containing catechol. Compared with the great progress in understanding catechol chemistry based on MAP in the last two decades, the chemistry of the barnacle cement protein and its adhesion mechanism are still mostly unknown, probably due to the inability to achieve characterization in a controlled way. To overcome this issue, much more fundamental research on various marine sessile organisms and close collaboration between biologists and engineers are necessary. On the other hand, catechol derivatives are widely found in not only mussels but also plants. For example, urushiol in Japanese lacquer, catechin, and tannin also have a catechol moiety and the potential to produce eco-friendly, tough coating materials, biomedical adhesives, self-healing surfaces, and antifouling surface modifiers. It is interesting to consider the abundance of novel material chemistry intimated in natural biological resources beyond our knowledge.

References

1. Waite, J. H. and Tanzer M. L. (1980). The bioadhesive of Mytilus byssus: a protein containig L-DOPA, *Biochem. Biophys. Res. Commun.*, **96**, pp. 1554–1561.
2. Hong, S., Kim, J., Na, Y. S., Park, J., Kim, S., Singha, K., Im, G., Han, D.-K., Kim, W. J. and Lee, H. (2013). Poly(norepinephrine): ultrasmooth material-independent surface chemistry and nanodepot for nitric oxide, *Angew. Chem. Int. Ed.*, **52**, pp. 9187–9191.
3. Li, K., Geng, X., Simonsen, J. and Karchesy, J. (2004). Novel wood adhesives from condensed tannins and polyethylenimine, *Int. J. Adhes. Adhes.*, **24**, pp. 327–333.
4. Watanabe, H., Fujimoto, A. and Takahara, A. (2014). Surface texturing of natural 'urushi' thermosetting polymer thin films, *Polym. J.*, **46**, pp. 216–219.

5. Watanabe, H., Fujimoto, A., Yamamoto, R., Nishida, J., Kobayashi, M. and Takahara, A. (2014). Scaffold for growing dense polymer brushes from a versatile substrate, *ACS Appl. Mater. Interfaces*, **6**, pp. 3648–3653.
6. Waite, J. H., Qin, X.-X. and Coyne, K. J. (1998). The peculiar collagens of mussel byssus, *Matrix Biol.*, **17**, pp. 93–106.
7. Waite J. H. (1983). Evidence for a repeating 3,4-dihydroxyphenylalanine- and hydroxyproline-containing decapeptide in the adhesive protein of the mussel, Mytilus edulis L., *J. Biol. Chem.*, **258**, pp. 2911–2915.
8. Waite, J. H. (2002). Adhesion à la Molule, *Inter. Comp. Biol.*, **42**, pp. 1172–1180.
9. Zhao, H. and Waite, J. H. (2006). Linking adhesive and structural proteins in the attachment plaque of Mytilus californianus, *J. Biol. Chem.*, **281**, pp. 26150 26158.
10. Lin, L. and Zeng, H. (2016). Marine mussel adhesion and bio-inspired wet adhesion, *Biotribology*, **5**, pp. 44–51.
11. Lee, H. (2006). Single-molecule mechanics of mussel adhesion, *Proc. Natl. Acad. Sci. U. S. A.*, **103**, pp. 12999–13003.
12. Ooka, A. A. and Garrell, R. L. (2000). Surface-enhanced Raman spectroscopy of DOPA-containing peptides related to adhesive protein of marine mussel, Mytilus edulis, *Biopolymers*, **57**, pp. 92–102.
13. Yu, M., Hwang, J. and Deming, T. J. (1999). Role of L-3,4-dihydroxyphenylalanine in mussel adhesive proteins, *J. Am. Chem. Soc.*, **121**, pp. 5825–5826.
14. Janković, I. A., Šaponjić, Z. V., Čomor, M. I. and Nedeljković, J. M. (2009). Surface modification of colloidal TiO2 nanoparticles with bidentate benzene derivatives, *J. Phys. Chem. C*, **113**, pp. 12645–12652.
15. Hwang, D. S., Harrington, M. J., Lu, Q., Masic, A., Zeng, H. and Waite, J. H. (2012). Mussel foot protein-1 (mcfp-1) interaction with titania surfaces, *J. Mater. Chem.*, **22**, pp. 15530–15533.
16. Hwang, D. S., Zeng, H., Masic, A., Harrington, M. J., Israelachvili, J. N. and Waite, J. H. (2010). Protein- and metal-dependent interactions of a prominent protein in mussel adhesive plaques, *J. Biol. Chem.*, **285**, pp. 25850–25858.
17. Yamamoto, H., Kuno, S., Nagai, A. Nishida, A., Yamauchi, S. and Ikeda, K. (1990). Insolubilizing and adhesive studies of water-soluble synthetic model proteins, *Int. J. Biol. Macromol.*, **12**, pp. 305–310.
18. Yu, M. and Deming, T. J. (1998). Synthetic polypeptide mimics of marine adhesives, *Macromolecules*, **31**, pp. 4739–4745.

19. Cha, H. J., Hwang, D. S., Lim S., White, J. D., Matos-Perez, C. R. and Wilker J. J. (2009). Bulk adhesive strength of recombinant hybrid mussel adhesive protein, *Biofouling*, **25**, pp. 99–107.

20. Lee, B. P., Dalsin, J. L. and Messersmith, P. B. (2002). Synthesis and gelation of DOPA-modified poly(ethylene glycol) hydrogels, *Biomacromolecules*, **3**, pp. 1038–1047.

21. Lee, B. P., Chao, C.-Y., Nunalee, F. N., Motan, E., Shull, K. R. and Messersmith, P. B. (2006). Rapid gel formation and adhesion in photocurable and biodegradable block copolymers with high DOPA content, *Macromolecules*, **39**, pp. 1740–1748.

22. Westwood, G., Horton, T. N. and Wilker, J. J. (2007). Simplified polymer mimics of cross-linking adhesive proteins, *Macromolecules*, **40**, pp. 3960–3964.

23. White, J. D. and Wilker, J. J. (2011). Underwater bonding with charged polymer mimics of marine mussel adhesive proteins, *Macromolecules*, **44**, pp. 5085–5088.

24. Nishida, J., Kobayashi, M. and Takahara, A. (2013). Light-triggered adhesion of water-soluble polymers with a caged catechol group, *ACS Macro Lett.*, **2**, pp. 112–115.

25. Nishida, J., Kobayashi, M. and Takahara, A. (2013). Gelation and adhesion behavior of mussel adhesive protein mimetic polymer, *J. Polym. Sci., Part A: Polym. Chem.*, **51**, pp. 1058–1065.

26. Lee, Y., Chung, H. J., Yeo, S., Ahn, C.-H. Lee, H., Messersmith, P. B. and Park, T. G. (2010). Thermo-sensitive, injectable, and tissue adhesive sol-gel transition hyaluronic acid/ pluronic composite hydrogels prepared from bio-inspired catechol-thiol reaction, *Soft Matter*, **6**, pp. 977–983.

27. Ryu, J. H., Lee, Y., Kong, W. H., Kim, T. G., Park, T. G. and Lee, H. (2011). Catechol-functionalized chitosan/pluronic hydrogels for tissue adhesives and hemostatic materials, *Biomacromolecules*, **12**, pp. 2653–2659.

28. Shin, M., Park, S.-G. Oh, B.-C., Kim, K., Jo, S., Lee, M. S., Oh, S. S., Hong, S.-H., Shin, E.-C., Kim, K.-S., Kang, S.-W. and Lee, H. (2016). Complete prevention of blood loss with self-sealing haemostatic needles, *Nat. Mater.*, **16**, pp. 147–154.

29. Harrington, M. J., Masic, A., Holten-Andersen, N., Waite, J. H. and Fratzl, P. (2010). Iron-clad fibers: a metal-based biological strategy for hard flexible coatings, *Science*, **328**, pp. 216–220.

30. Wilker, J. J. (2010). The iron-fortified adhesive system of marine mussels, *Angew. Chem., Int. Ed.*, **49**, pp. 8076–8078.

31. Avdeef, A., Sofen, S. R., Bregante, T. L. and Raymond, K. N. (1978). Coordination chemistry of microbial iron transport compounds. 9. Stability constants for catechol models of enterobactin, *J. Am. Chem. Soc.*, **100**, pp. 5362–5370.

32. Sever, M. J. and Wilker, J. J. (2004). Visible absorption spectra of metal-catecholate and metal-tironate complexes, *Dalton Trans.*, pp. 1061–1072.

33. Xu, H. Nishida, J., Ma, W. Wu, H., Kobayashi, M. Otsuka, H. and Takahara, A. (2012). Competition between oxidation and coordination in cross-linking of polystyrene copolymer containing catechol groups, *ACS Macro Lett.*, **1**, pp. 457–460.

34. Holten-Andersen, N., Harrington, M. J., Birkedal, H., Lee, B. P., Messersmith, P. B., Lee, K. Y. C. and Waite, J. H. (2011). pH-induced metal-ligand cross-links inspired by mussel yield self healing polymer networks with near-covalent elastic moduli, *Proc. Natl. Acad. Sci. U.S.A.*, **108**, pp. 2651–2655.

35. He, L. H., Fullenkamp, D. E., Rivera, J. G. and Messersmith, P. B. (2011). pH responsive self-healing hydrogels formed by boronate-catechol complexation, *Chem. Commun.*, **47**, pp. 7497–7499.

36. Lee, H., Dellatore, S. M., Miller, W. M. and Messersmith, P. B. (2007). Mussel-inspired surface chemistry for multifunctional coatings, *Science*, **318**, pp. 426–430.

37. Ye, Q., Zhou, F. and Liu, W. (2011). Bioinspired catecholic chemistry for surface modification, *Chem. Soc. Rev.*, **40**, pp. 4244–4258.

38. Dalsin, J. L., Hu, B.-H., Lee, B. P. and Messersmith, P. B. (2003). Mussel adhesive protein mimetic polymers for the preparation of nonfouling surfaces, *J. Am. Chem. Soc.*, **125**, pp. 4253–4258.

39. Lee, H., Lee, B. P. and Messersmith, P. B. (2007). A reversible wet/dry adhesive inspired by mussels and geckos, *Nature*, **448**, pp. 338–342.

40. You, I., Kang, S. M., Byun, Y. and Lee, H. (2011). Enhancement of blood compatibility of poly(urethane) substrates by mussel-inspired adhesive heparin coating, *Bioconjugate Chem.*, **22**, pp. 1264–1269.

41. Fan, X., Lin, L., Dalsin, J. L. and Messersmith, P. B. (2005). Biomimetic anchor for surface-initiated polymerization from metal substrates, *J. Am. Chem. Soc.*, **127**, pp. 15843–15847.

42. Hong, S. H., Hong, S., Ryou, M.-H., Choi, J. W., Knag, S. M. and Lee, H. (2016). Sprayable ultrafast polydopamine surface modifications, *Adv. Mater. Interfaces*, **3**, p. 15000857.

43. Kimura, S., Ohshima, C., Hirose, E., Nishikawa, J. and Itoh, T. (2001). Cellulose in the house of the appendicularian Oikopleura rufescens, *Protoplasma*, **216**, pp. 71–74.

44. Capadona, J. R., Shanmuganathan, K., Tyler, D. J., Rowan, S. J. and Weder, C. (2008). Stimuliresponsive polymer nanocomposites inspired by the sea cucumber dermis, *Science*, **319**, pp. 1370–1374.

45. Taylor, S. W., Kammerer, B. and Bayer, E. (1997). New perspectives in the chemistry and biochemistry of the tunichromes and related compounds, *Chem. Rev.*, **97**, pp. 333–346.

46. Bayer, E., Schiefer, G., Waidelich, D., Scippa, S. and De Vicentiis, M. (1992). Structure of the tunichrome of tunicates and its role in concentrating vanadium, *Angew. Chem. Int. Ed.*, **31**, pp. 52–54.

47. Kime-Hunt, E., Spartalian, K., Holmes, S., Mohan, M. and Carrano, C. J. (1991). Vanadium metabolism in tunicates: the coordination chemistry of V(III), V(IV), and V(V) with models for the tunichromes, *J. Inorg. Biochem.*, **41**, pp. 125–141.

48. Cai, M., Sugumaran, M. and Robinson, W. E. (2008). The crosslinking and antimicrobial properties of tunichrome, *Comp. Biochem. Physiol. B: Biochem. Mol. Biol.*, **151**, pp. 110–117.

49. Oh, D. X., Kim, S., Lee, D. and Hwang, D. S. (2015). Tunicate-mimetic nanofibrous hydrogel adhesive with improved wet adhesion, *Acta Biomater.*, **20**, pp. 104–112.

50. Walker, G. (1972). The biochemical composition of the cement of two barnacle species, Balanus hameri and Balanus crenatus, *J. Mar. Biol. Assoc. U. K.*, **52**, pp. 429–435.

51. Kamino, K. (2008). Underwater adhesive of marine organisms as the vial link between biological science and material science, *Mar. Biotechnol.*, **10**, pp. 111–121.

52. Kamino, K., Inoue, K., Maruyama,T., Takamatsu, N., Harayama, S. and Shizuri, Y. (2000). Barnacle cement proteins: importance of disulfide bonds in their insolubility, *J. Biol. Chem.*, **275**, pp. 27360–27365.

53. Kamino, K. (2010). Absence of cross-linking via trans-glutaminase in barnacle cement and redefinition of the cement, *Biofouling*, **26**, pp. 755–760.

54. Kamino, K. (2010). Molecular design of barnacle cement in comparison with those of mussel and tubeworm, *J. Adhes.*, **86**, pp. 96–110.

55. Urushida, Y., Nakano, M., Matsuda, S., Inoue, N., Kanai, S., Kitamura, N., Nishino, T. and Kamino, K. (2007). Identification and functional characterization of a novel barnacle cement protein, *FEBS J.*, **274**, pp. 4336–4346.

56. Nishida, J. Higaki, Y. and Takahara, A. (2015). Synthesis and characterization of barnacle adhesive mimetic towards underwater adhesion, *Chem. Lett.*, **44**, 1047–1049.

Chapter 13

Functional Analysis of the Mechanical Design of the Cricket's Wind Receptor Hair

Tateo Shimozawa
Hokkaido University, Sapporo, Hokkaido 060-0808, Japan
tsmz@es.hokudai.ac.jp

The intrinsic mechanical parameters of a cricket's cercal wind receptor hair were determined from direct measurement of hair deflections. The tangential velocity of the deflecting hair shaft in response to stimulus air motion was measured in situ by a laser-Doppler velocimeter with surface scattering of the shaft. The velocity of the stimulus air motion was calibrated by the same velocimeter with smoke particle scattering. The mobility of the hair was obtained from the former measurement with reference to the latter calibration of the single apparatus. A Gaussian white noise (GWN) signal was employed as the stimulus waveform, and the stimulus-response transfer function was calculated through a cross correlation method, which provides greater precision and wider frequency for a longer period of measurement. The hairs were modeled as an inverted pendulum, and the mechanical parameters

Industrial Biomimetics
Edited by Akihiro Miyauchi and Masatsugu Shimomura

ISBN 978-981-4800-07-5 (Hardcover), 978-0-429-05883-7 (eBook)
www.jennystanford.com

of the equation of motion were determined from data given by the systematic measurement of mobility by the least-squares error sum method. The theoretical torque that deflects the hair shaft is given by the viscous force calculated from Stokes's mechanical impedance of an oscillating cylinder in a viscous fluid. In addition to the effect of the boundary layer in which air is stagnating on the substrate surface, hair diameter is taken into account for viscous force calculation. Through these analyses, a structural scaling in the biological design of the mechanical interface for the sensory systems is revealed.

13.1 Introduction

Crickets and cockroaches have a pair of appendages called cerci at the rear end of the abdomen (Fig. 13.1A). On the cercus, thin filiform hairs 1–10 μm in basal diameter and 30–1500 μm in length protrude into the air (Fig. 13.1B). The filiform hairs observed under a dissection microscope show deflections when the air moves. A sensory cell located at the hair base receives a mechanical pinch on the tip of the sensory cilium [5, 11]. In response to the mechanical stimulus, a molecular transduction mechanism of the sensory cell generates a membrane potential and transmits a pulse train to the central nervous system (CNS) (Fig. 13.1C). The neuronal pulse is conducted along the axon of the sensory cell and transfers signal to interneurons in the CNS. Signals from a variety of sensory cells represent information of air motion in the external world through synaptic integration and finally exert a behavioral output such as an evasion response against a predatory attack. A filiform hair protruding into the air is a mechanical interface that mediates air motion to the sensory cell and is called a wind receptor hair. Air motion–sensitive animals can detect a moving object nearby without relying on vision. The biological significance and sensitivity of the cricket wind receptor system have been introduced and discussed elsewhere [11, 12, 15, 16]. Here the author explains the way of using laser-Doppler velocimetry (LDV) on the small-sized natural gift machinery, the method to analyze the mechanical design by using Stokes's theory of viscosity, and the biological results obtained [8, 9, 17].

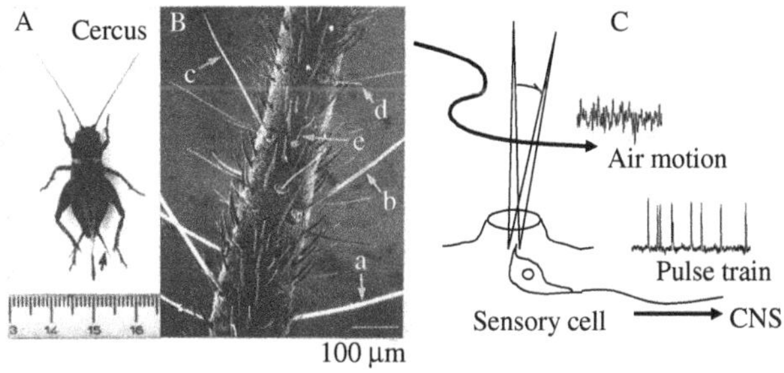

Figure 13.1 (A) Cercus of the field black cricket, *Gryllus bimaculatus* (arrow). (B) SEM micrograph of the cercal surface; (a–e) Size variety of the filiform hair (corresponding to Fig. 13.2B). (C) Schematic representation of the wind sensory system; deflection of the hair due to air motion elicits a neuronal pulse train propagating to the CNS.

13.2 Measurement of Mobility of the Hair Due To Air Motion

Filiform hairs of a cricket normally respond to air motions 1–100 Hz in frequency and 0.1–100 mm/s in air velocity (Fig. 13.2A). Although they also respond to sound (propagating pressure wave), the pressure difference between frontal and rear surfaces of the hair is not the driving force because the hair is so thin in diameter (Fig. 13.2B). The Reynolds number for the hair diameter and air velocity is 10^{-2} at most. Therefore, deflections of sensory hairs are measured and analyzed as a mechanical system driven by a viscous force resulting from air motion.

First, the tangential motion velocity of the hair shaft was measured by a laser-Doppler velocimeter (LDV: DANTEC 55N21) (Fig. 13.3). The intersection of two laser beams, one a 514.5 nm reference beam and the other a 40 MHz frequency-shifted 514.5 nm beam, by an optoacoustic Doppler device, with a cross angle of 13.5° produces a moving interference fringe pattern with spatial period 2.186 μm and moving speed 87.4 m/s. The intensity of the back-scattered light from a stationary particle or a point on the rough surface within the moving fringe varies with 40 MHz frequency.

The motion of the light-scattering particles or surface in the same (or opposite) direction as the fringe motion causes a reduction (or increase) in the frequency of the scattered light. Next, the frequency of the back-scattered light collected by a photomultiplier tube was converted into the output signal voltage, which is proportional to the instantaneous velocity of the scattering sources within the measurement volume. The size of the measurement volume is 75 μm in terms of cross-sectional diameter and 600 μm in length along the beam. Back scattering was good enough for hair velocity measurement. The stimulus air motion was generated in a wind tunnel consisting of two loudspeakers attached to both ends of an acrylic tube (30 mm inner diameter, 150 mm length). The loud speakers were driven by a power amplifier in a push-pull fashion along the axis of the tube and produced an alternating air motion along the wind tunnel. The velocity of the air motion stimulus in the wind tunnel was calibrated with cooled smoke from a joss stick (Fig. 13.4). Smoke particles suspended in the moving air mass scatter the light.

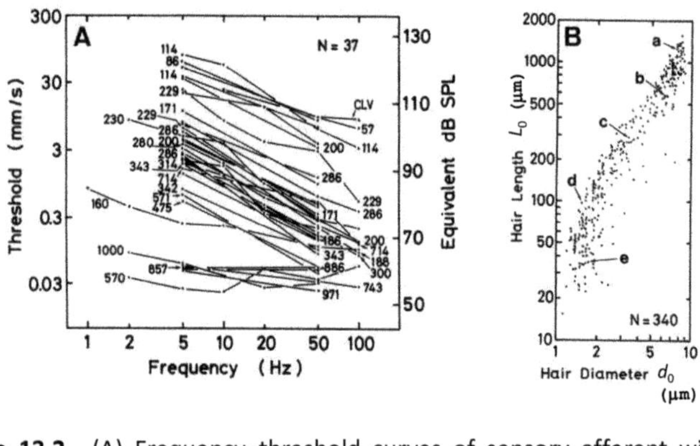

Figure 13.2 (A) Frequency–threshold curves of sensory afferent with hair length (numbers in μm). Abscissa: alternating frequency of stimulus air motion. Ordinate: threshold value in peak velocity of sinusoidal air motion (left). Note that long hairs are velocity sensitive and short ones are acceleration sensitive. (B) Size distribution of filiform hair. Abscissa: hair diameter at the base. Ordinate: hair length. (Arrows a–e) Hairs corresponding to those marked with letters in Fig. 13.1B.

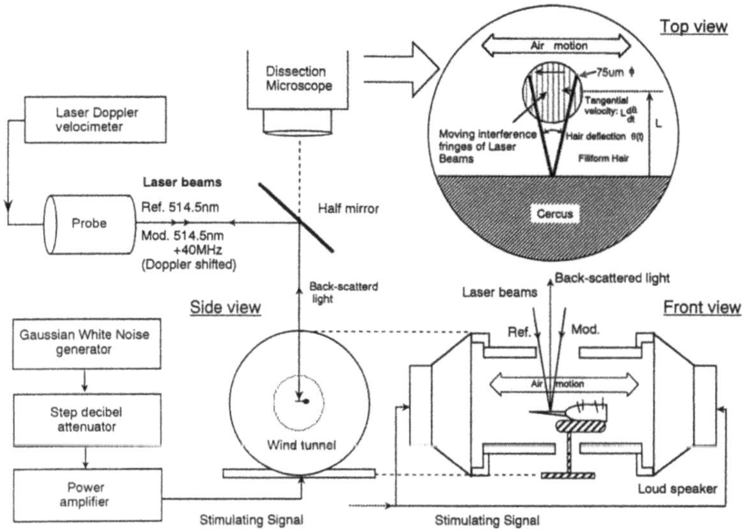

Figure 13.3 Setup for measurements of both velocity of the stimulus air motion and tangential velocity of filiform hairs. The intersection of reference and Doppler-shifted beams of LDV was focused on a single hair on the cercus through the half mirror. The stimulus air motion was modulated by GWN.

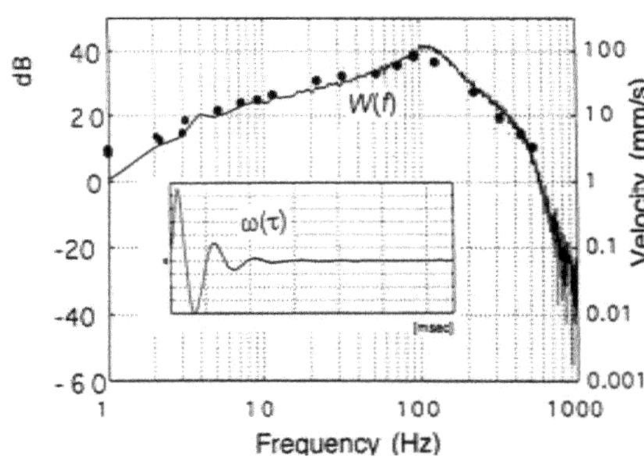

Figure 13.4 Wind tunnel calibration: electromechanical transfer function from driving signal to air motion velocity. Inset: Impulse response $\omega(\tau)$ obtained by the cross correlation method with a 1 Vrms GWN. $W(f)$: frequency characteristics. Solid line: Fourier transform of $\omega(\tau)$ (ordinate at the left). Dots: Peak air velocities to the sinusoid of 1 Vpeak at various frequencies obtained by stimulus synchronized summation average (ordinate at the right).

Because the hair is small in length (Fig. 13.2B), the tangential velocity signal is very small and buried under LDV noise. To improve the signal-to-noise ratio (S/N), the transfer function from the system input to a measured output was obtained by a statistical method utilizing a GWN signal. The cross correlation function between the GWN input $x(t)$ and the measured output $y(t)$, calculated for an appropriately long observation time T_m, gives the impulse response $h(\tau)$, which uniquely characterizes a linear system [10].

$$h(\tau)=\frac{1}{P}\lim_{T_m\to\infty}\frac{1}{T_m}\int_0^{T_m} y(t)\cdot x(t-\tau)dt\,, \tag{13.1}$$

where P is the power spectrum density of the input GWN. A Fourier transform of $h(\tau)$ by a conventional fast Fourier transform (FFT) gives the frequency-domain transfer function $H(f)$ of the system. The use of the GWN, a random signal with a flat power-spectral density for all frequencies with an amplitude probability function of Gaussian distribution whose autocorrelation is a delta function, provides a convenient statistical way to obtain the linear transfer function. The method has the merit of high tolerance against any internal noise made by the system, including the measurement apparatus. Internal and digitizing noises added linearly to the output are canceled out by correlation with a stimulus GWN input independent of the noises. With a longer observation time T_m, a better estimation can be given. For each hair, the impulse response of 2048 point for 1024 ms was determined from a pair of input-output data of 300 s. The GWN signal was obtained from the low-pass filtering of a pseudorandom binary m-sequence generated by a 42-bit shift register sequence (NF CG-742N) driven by a 25 kHz clock. The frequency bandwidth of the GWN was 500 Hz, 50 Hz, or 20 Hz.

Tangential velocity (Fig. 13.5) was converted into angular velocity of hair deflection by dividing it by height from the hair base to measurement volume. Time domain integration of the angular velocity gives hair deflection (Fig. 13.6). Because the velocities of stimulus air and hair shaft were measured by the same instrument, the ratio of transfer functions for hair deflection relative to that for stimulus air velocity represents the mobility (mechanical characteristics) of each hair. In some cases, the measurement time of 300 s is not good enough to overcome existing instrumental and digitizing noises at a lower-frequency range. To improve S/N at a lower frequency of the hair shaft velocity, the bandwidth of the GWN was narrowed while keeping its root mean square amplitude constant. The band narrowing from 500 Hz to 20 Hz results in a 25-fold increase in the power spectrum density of the driving GWN. This is equivalent to an increase in the measurement period to 7500 s (2 h). Compared with discrete measurements by sinusoid stimulus, the improvement of the S/N by the narrow-band GWN was confirmed (Fig. 13.5 dots and dashed line). As a further criterion for the validity of measurement, the coherence function between the stimulus GWN signal and the measured velocity signal was calculated. The coherence function is a cross product of covariance between the stimulus and response normalized by the sum of autocorrelations of the two. Only the measurements whose coherence was higher than 0.01 were used for further estimations of hair mobility.

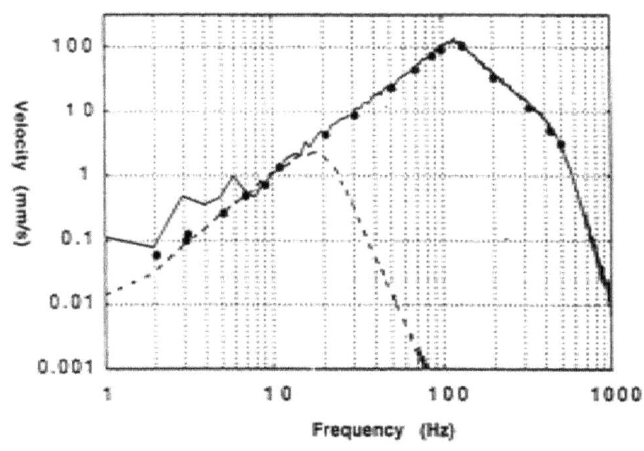

Figure 13.5 Frequency spectrum of transfer function of hair shaft velocity from the GWN signal. Solid line: Spectrum of impulse response obtained by 500 Hz 1 Vrms GWN for 300 s. Dashed line: S/N-improved transfer function spectrum obtained by 20 Hz 1 Vrms GWN for 300 s for the same hair shaft. Filled dots: Discrete measurements of hair shaft velocities in response to a sinusoid stimulus of 1 Vpeak at various frequencies.

The mobility of hair, in terms of deflection sensitivity against sinusoidal air motion at 1 mm/s in peak velocity, was measured for total of 27 hairs whose length ranged from 160 to 1500 μm. All hairs were confirmed to be mechanical bandpass filters with a length-dependent best frequency of 100–300 Hz in 160–600 μm hairs (Fig. 13.7-A1) and 30–60 Hz in 1000–1500 μm hairs (Fig. 13.7-B1). The best frequencies varied inversely to hair length: 160–600 μm hairs at 100–200 Hz, 600–900 μm hairs at 50–80 Hz, and 1000–1500 μm hairs at 30–60 Hz (Figs. 13.7-A1, 13.7-B1; Table 13.1). At the best frequencies, all hairs deflect roughly 10^{-3} [radians] in air motion of 1 mm/s at peak velocity. The filiform hairs cover a wide range of frequencies (10–500 Hz) with flat sensitivity as an array of bandpass filters of different lengths (30–1500 μm). All of the measured slopes halfway through the phase characteristics were rather moderate, about $\pi/2$ radians per decade change of frequency, regardless of the length of hair (Figs. 13.7-A2, 13.7-B2). This phase slope indicates that all hairs, irrespective of the length, are strongly damped second order systems whose damping factors are about 1.46, twice as large as that of the optimal damping.

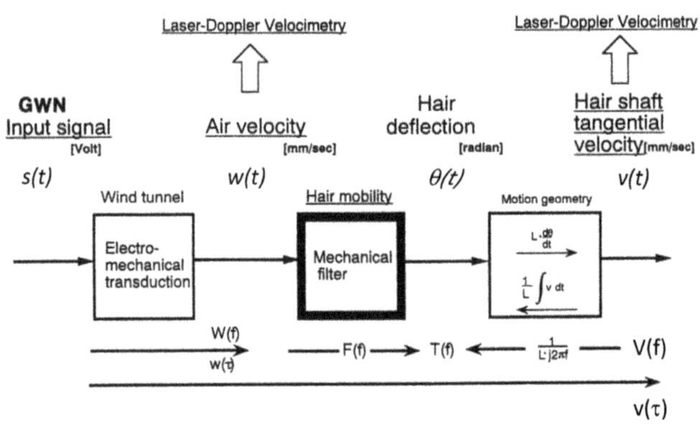

Figure 13.6 Relationship between physical quantities in the air-hair system and transfer functions in the data processing. Upper rows: Physical quantities and units of measurement; the actually measured quantities are underlined. Blocks in the middle row: Physical transductions at each mechanical element. Arrows connecting the blocks: Physical causalities. Arrows in bottom rows: Frequency transfer functions and equivalent time-domain impulse responses obtained by the cross correlation method with the GWN (*W*(*f*), *V*(*f*)) and those estimated through calculation (*T*(*f*), *F*(*f*)). Leftward arrow: Frequency-domain determination of hair deflection from tangential velocity.

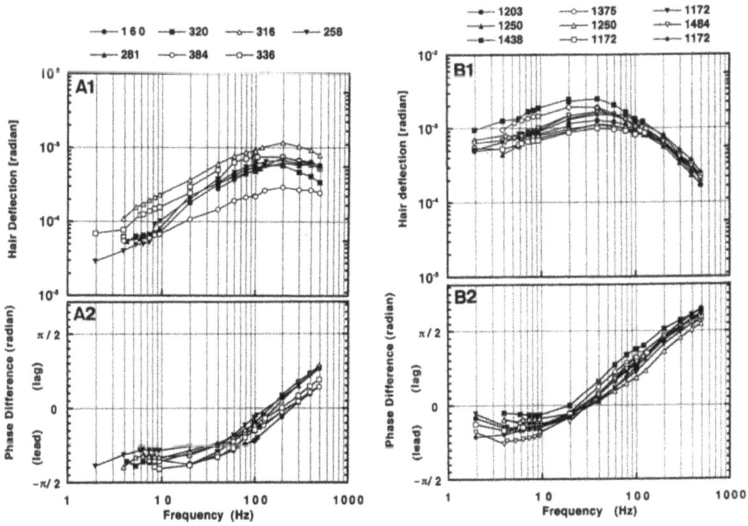

Figure 13.7 The Bode diagrams of the deflection amplitude for hairs shorter than 400 μm (A) and hairs longer than 1000 μm (B). Abscissa: log-scaled frequency. (A1) Gain characteristics of short hairs; the low-frequency side slope is 15 dB/decade. (A2) Phase characteristics of short hairs; $\pi/4$ phase advance in lower frequencies with $\pi/2$ per decade slope in higher frequencies. (B1 and B2) Gain and phase for long hairs, both shifted to lower frequencies compared to those of short hairs.

13.3 Mechanical Model of the Wind Receptor Hair

The cercal wind receptor hair of the cricket is modeled as an inverted pendulum with a cone-shaped solid shaft supported by a spring at the base (Fig. 13.8). The pendulum shaft that protrudes into the air is driven by the viscous force applied by the moving air. The inverted pendulum is a typical second order mechanical system that consists of a moment of inertia that represents the mass distribution along the hair shaft, a spring that provides the restoring torque toward a resting position of the hair, and torsional resistance within the hair base. The angular deflection $\theta(t)$ satisfies the fundamental equation of motion of the pendulum at every instance in time. For the conservation of angular momentum,

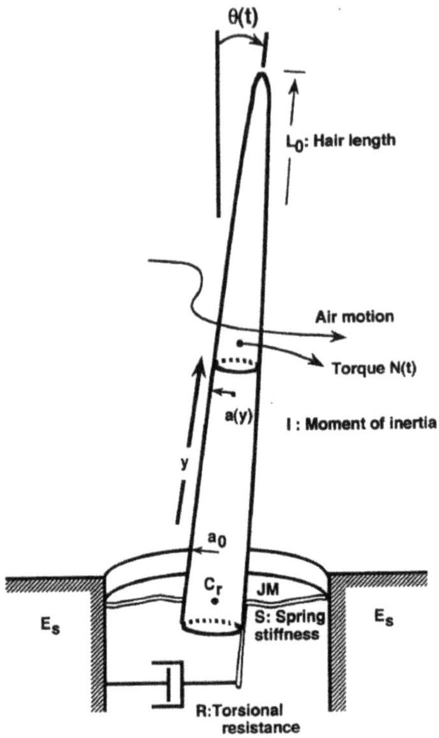

Figure 13.8 Schematic representation of the mechanical structure of a wind receptor hair. The hair shaft, with an inertia moment I, is supported by a circular diaphragm of the joint membrane (JM) with a spring stiffness S. The hair is driven by a torque $N(t)$ due to drag force from moving air and causes an angular deflection $\theta(t)$. The mechanical energy of the hair motion is consumed by a resistance R of wet tissue within the hair base. (y, distance variable along the hair shaft from the base, the arm length of torque generation; $a(y)$, radius of the hair at y; L_0, a_0, length and base radius of the hair, respectively; E_s, exoskeletal substrate; and C_r, center of rotation for hair motion.)

$$I \cdot \frac{d^2\theta(t)}{dt^2} + R \cdot \frac{d\theta(t)}{dt} + S \cdot \theta(t) = N(t)\,,\ [\mathrm{N\ m}] \tag{13.2}$$

where I is the moment of inertia of the hair shaft, R is the torsional resistance within the hair base, S is the spring stiffness of the hair supporting base, and $N(t)$ is the external torque acting on the hair shaft due to the viscous force from moving air. As shown in the previous section, $\theta(t)$ of a wind receptor hair to the velocity sinusoid is measurable. Therefore, we can determine three intrinsic mechanical parameters I, R, and S of the hair if $N(t)$ is theoretically given. Since hairs left under a measurement apparatus for a long time become immobile—less than 1/20 in deflection amplitude probably because the cercus had dried up—the pendulum shaft is rigid and the viscous force from moving air does not bend it.

We have a set of data of hair mobility, pairs of deflection amplitude Θ_m and phase shift φ_m in response to the sinusoidal stimulus air motion of known velocity (1 mm/s) at several frequencies f_m. By using the pair of values, we can express the hair deflection at time t_s as

$$\theta(t_s) = \Theta_m \cdot \sin(2\pi\phi_m \cdot t_s + \varphi_m). \qquad \text{[radians]} \quad (13.3\text{-}1).$$

Therefore,

$$\frac{d\theta(t_s)}{dt} = (2\pi f_m) \cdot \Theta_m \cdot \cos(2\pi f_m \cdot t_s + \varphi_m), \quad [\text{radians s}^{-1}] \quad (13.3\text{-}2)$$

and

$$\frac{d^2\theta(t_s)}{dt^2} = -(2\pi f_m)^2 \cdot \Theta_m \cdot \sin(2\pi f_m \cdot t_s +_m). \quad [\text{radians s}^{-2}] \; (13.3\text{-}3)$$

We calculated the numerical values of these quantities by assigning four instances of time in a cycle of the stimulus sinusoid, $t_s = s/(4f_m)$, (s = 0, 1, 2, and 3). By giving the theoretical values of the external torque $N(t_s)$ at each instance of time (see the next section for torque calculation), we obtained four simultaneous equations for the three unknowns, I, R, and S. This assignment of sampling time is redundant but was employed just for the sake of simplicity and caused no excess error in the least-squares error sum estimation. The mobility data of a single hair is a set of Θ_m, φ_m at 14–18 different frequencies (f_m) ranging from 2 to 500 Hz. By replacing all time variables with the numerical values of Θ_m, φ_m, and f_m using Eqs. 13.3-1 to 13.3-3, we obtained 56–72 simultaneous equations for the three unknowns of a single hair, although half of these were redundant. The three unknown parameters were determined so as to satisfy the large number of equations with the minimum square error sum [8, 17]. See the appendix (at the end of this chapter) for details.

13.4 Stokes's Mechanical Impedance of an Oscillating Cylinder in a Viscous Fluid

Stokes [19] gave a rigorous linear solution of the Navier–Stokes equation of fluid dynamics for the amount of drag force caused by a cylinder moving in a viscous fluid with a velocity sinusoid. The solution was originally derived to consider the effect of air viscosity on the swinging period of a pendulum, which was at that time used for the precise determination of the gravity acceleration of the earth [19]. We may apply the solution to essentially the same problem for the force generated on the shaft of the inverted pendulum driven by sinusoidal air motion.

When a circular cylinder of radius a (=$d/2$) is moving in an infinitely-large viscous fluid back and forth in the direction perpendicular to the cylinder axis with the velocity,

$$u(t) = U_0 \cdot \sin(2\pi f \cdot t),\ [\mathrm{m\ s^{-1}}]$$

the drag force generated on the cylinder of unit length is given by

$$F_s(t) = \left\{ 4\pi\mu \cdot \frac{-g}{g^2 + (\pi/4)^2} \right\} U_0 \cdot \sin(2\pi f \cdot t)$$

$$+2\pi f \left\{ \pi\rho\ a^2 + \frac{2\mu}{f} \cdot \frac{\pi/4}{g^2 + (\pi/4)^2} \right\} U_0 \cdot \cos(2\pi f \cdot t),\ [\mathrm{N\ m^{-1}}] \quad (13.4)$$

where $g = \ln(s) + \gamma$, $s = (a/2)\sqrt{2\pi f/\nu}$, U_0 is the peak velocity of the sinusoidal motion of the cylinder, ρ is the density of air (1.205 kg m^{-3} at 20°C), μ is the viscosity of air (1.82×10^{-5} N s m^{-2} at 20°C), ν is the kinematic viscosity of air (1.51×10^{-5} m^2 s^{-1} at 20°C), and γ is Euler's constant (=0.57721). Because this is a linear solution of the equation, the relation between force and velocity can be expressed by a conversion factor from one to the other and is referred to as mechanical impedance. Stokes's mechanical impedance for an oscillating cylinder is a complex number corresponding to the presence of the in-phase force in addition to the $\pi/2$ phase-shifted component. By using the notation of a complex number, Stokes's mechanical impedance Z_S is expressed as

$$Z_S = Z_{SR} + j \cdot Z_{SX},\quad [\mathrm{N\ s\ m^{-2}}] \quad (13.5)$$

where j is the $\pi/2$ phase rotator or $j^2 = -1$,

$$Z_{SR} = 4\pi\mu \cdot \frac{-g}{g^2 + (\pi/4)^2}, \text{ and } Z_{SX} = 2\pi f \left\{ \pi\rho a^2 + \frac{2\mu}{f} \cdot \frac{\pi/4}{g^2 + (\pi/4)^2} \right\}. \tag{13.6}$$

With this notation for the real and imaginary parts of the impedance, the dynamic drag force caused by the velocity sinusoid is expressed as

$$F_S(t) = U_0 \cdot |Z_S| \sin(2\pi f \cdot t + \eta_S), \tag{13.7}$$

where $|Z_S| = \sqrt{Z_{SR}^2 + Z_{SX}^2}$ and $\eta_S = \tan^{-1}(Z_{SX}/Z_{SR})$.

These are the magnitude and the phase shift of the theoretical drag force.

13.5 Boundary Layer

The velocity of air motion in the vicinity of the cercal surface is attenuated because of the viscous stagnation of the air around the substrate. The air mass that stagnates on the substrate is called the boundary layer. The theory of air motion produced by a flat and infinitely large plate oscillating parallel to the surface is known as Stokes's second problem [6, 19]. By a slight modification of the coordinate system of this theory, we can derive the distribution of air velocity near the substrate. The velocity of air motion near the surface of the infinite flat substrate is given by the following formula when air moves parallel to the surface:

$$u_y(t) = U_0 \sin(2\pi f \cdot t) - U_0 \cdot e^{-\beta \cdot y} \cdot \sin(2\pi f \cdot t - \beta \cdot y), \text{ [m s}^{-1}\text{]} \tag{13.8}$$

where $U_0 \sin(2\pi f \cdot t)$ is the velocity of air motion at a distance infinitely far from the substrate, y is the distance from substrate surface, β (= $\sqrt{\pi f/v}$) is the reciprocal of the boundary layer thickness, f is the frequency of air motion, and v is the kinematic viscosity of air. In lower frequencies, the boundary layer becomes thicker with $f^{-1/2}$ and the air near the substrate stagnates (Fig. 13.9). Short hairs receive no air motion in lower frequencies. To express the boundary layer effect at a height y, Eq. 13.7 is transformed to express in amplitude and phase shift of the sinusoid:

$$u_y(t) = U_y \sin(2\pi f \cdot t + \xi_y), \text{ [m s}^{-1}\text{]} \tag{13.9}$$

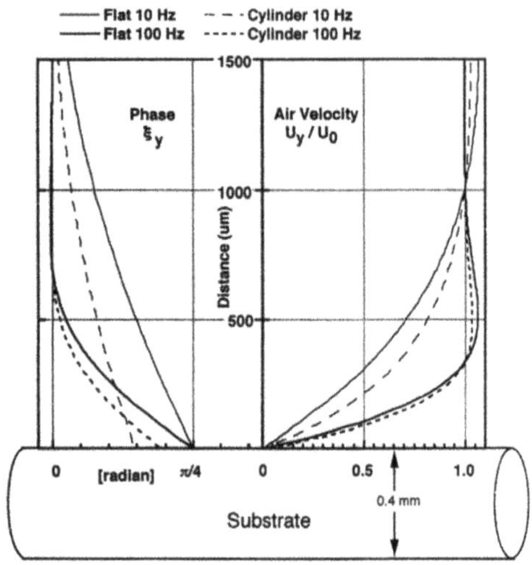

Figure 13.9 Profiles of air motions in the boundary layer as functions of distance from the substrate surface (ordinate). Velocity amplitudes (right) and phase shifts (left) of air motion for infinite flat (solid curves) and cylindrical (dashed curves) substrates are compared at two frequencies (10 and 100 Hz). Air motions near the surface stagnate; the air velocity is damped, and its phase advances to $\pi/4$ (radians) compared to that at an infinite distance. The boundary layers on cylindrical substrates have similar forms to those on the infinite flat one.

where $U_y = U_0\sqrt{\left(1-e^{-\beta\cdot y}\cdot\cos(\beta\cdot y)\right)^2+\left(e^{-\beta\cdot y}\cdot\sin(\beta\cdot y)\right)^2}$,

and $\xi_y = \tan^{-1}\left\{e^{-\beta\cdot y}\cdot\sin(\beta\cdot y)/\left(1-e^{-\beta\cdot y}\cdot\cos(\beta\cdot y)\right)\right\}$.

The drag force $F_{Sy}(t)$ generated on the hair shaft at the height y is given by the relative velocity $u_{ry}(t)$ and Stokes's mechanical impedance Z_{Sy} of the shaft. The relative velocity can be expressed as

$$\begin{aligned} u_{ry}(t) &= U_y\sin(2\pi f\cdot t+\xi_y)-y\cdot d\theta/dt \\ &= U_y\sin(2\pi f\cdot t+\xi_y)-y\cdot\Theta_m\cdot 2\pi f\cdot\cos(2\pi f\cdot t+\varphi_m), [\mathrm{m\ s^{-1}}] \end{aligned} \tag{13.10}$$

where U_y and ξ_y are the velocity amplitude and the phase shift of air motion at height y from the substrate surface and Θ_m and φ_m are the deflection amplitude and phase shift of the hair motion obtained by the mobility data. The drag force generated due to the mechanical impedance by this relative velocity is thus,

$$F_{Sy}(t) = |Z_{Sy}|\{U_y \sin(2\pi f \cdot t + \xi_y + \eta_{Sy})$$

$$-y \cdot \Theta_m \cdot 2\pi f \cdot \cos(2\pi f \cdot t + \phi_m + \eta_{Sy})\}, \tag{13.11}$$

where Z_{Sy} and η_{Sy} are the modulus and phase shift of Stokes's mechanical impedance of the hair shaft for the diameter at height y at frequency f.

13.6 Torque Calculation

The total torque $N(t)$ that turns the hair shaft around its base is given by the integrated sum of the small torque, each generated on an infinitesimally thin disc of hair shaft at the arm length of rotation y. Thus,

$$N(t) = \int_0^{L_0} F(y,t) \cdot y \cdot dy, \text{ [N m]} \tag{13.12}$$

where $F(y, t)$ represents the drag force per unit length acting on the hair shaft at a height y at a time t, and L_0 is the hair length. The drag force is a function of the shaft radius and the velocity of air relative to the shaft. By replacing $F(y, t)$ with Eq. 13.10, which represents the instantaneous drag force at y under a sinusoidal air motion, we get

$$N(t) = \int_0^{L_0} F_{Sy}(t) \cdot y \cdot dy$$

$$= \int_0^{L_0} \Big[U_y \cdot |Z_{Sy}| \cdot y \cdot \sin(2\pi f \cdot t + \xi_y + \eta_{Sy})$$

$$-y^2 \cdot 2\pi f \cdot \Theta_m \cdot |Z_{Sy}| \cos(2\pi f \cdot t + \varphi_m + \eta_{Sy})\Big] dy$$

$$= \left[\int_0^{L_0} \{U_y \cdot |Z_{Sy}| \cdot y \cdot \cos(\xi_y + \eta_{Sy} - \varphi_m)\right.$$

$$\left. + y^2 \cdot 2\pi f \cdot \Theta_m \cdot |Z_{Sy}| \cdot \sin(\eta_{Sy})\} dy\right] \cdot \sin(2\pi f \cdot t + \varphi_m)$$

$$+ \left[\int_0^{L_0} \{U_y \cdot |Z_{Sy}| \cdot y \cdot \sin(\xi_y + \eta_{Sy} - \varphi_m)\right.$$

$$\left. -y^2 \cdot 2\pi f \cdot \Theta_m \cdot |Z_{Sy}| \cdot \cos(\eta_{Sy})\} dy\right] \cdot \cos(2\pi f \cdot t + \phi_m). \tag{13.13}$$

Because the hair is rigid, or free from bending, Θ_m and φ_m are y-independent quantities when measured in a single hair at a specific frequency. So they can be placed outside the integrals. Therefore,

$$N(t)=\left[A\cdot\cos(\varphi_m)+B\cdot\sin(\varphi_m)+C\cdot 2\pi f\cdot\Theta_m\right]\cdot\sin(2\pi f\cdot t+\varphi_m)$$

$$+\left[B\cdot\cos(\varphi_m)-A\cdot\sin(\varphi_m)-D\cdot 2\pi f\cdot\Theta_m\right]\cdot\cos(2\pi f\cdot t+\varphi_m),\quad (13.14)$$

where

$$A=\int_0^{L_0} U_y\cdot\left|Z_{Sy}\right|\cdot y\cdot\cos(\xi_y+\eta_{Sy})dy,$$

$$B=\int_0^{L_0} U_y\cdot\left|Z_{Sy}\right|\cdot y\cdot\sin(\xi_y+\eta_{Sy})dy,$$

$$C=\int_0^{L_0}\left|Z_{Sy}\right|\cdot y^2\cdot\sin(\eta_{Sy})dy=\int_0^{L_0} y^2\cdot\frac{\pi^2\mu}{g^2+(\pi/4)^2}dy,$$

and

$$D=\int_0^{L_0}\left|Z_{Sy}\right|\cdot y^2\cdot\cos(\eta_{Sy})dy=\int_0^{L_0} y^2\cdot 4\pi\mu\cdot\frac{-g}{g^2+(\pi/4)^2}dy.\quad (13.15)$$

Parameters A, B, C, and D are uniquely enumerated for each hair by the y dependencies both of air motion (i.e., the boundary layer effect) and of the mechanical impedance of the hair shaft (diameter). The shaft radius as a function of distance from the hair base (i.e., hair shape) is given below. We will be able to give values of total torque $N(t)$ at any instance of time if the length and mobility of a hair were measured. Numerical integrations along hair length for A, B, C, and D above were performed by Simpson's formula of approximation.

13.7 The Shape of the Wind Receptor Hair

For a correct analysis, accurate estimation of viscous force and turning torque is necessary. The main attribute that determines the amount of viscous drag force generated by moving air is the diameter of the shaft. The turning torque working on a single hair is an integrated sum of an infinitesimally small torque that is itself the product of the drag force and the arm length from the hair base. In addition to the torque, the moment of inertia of a hair shaft is also the sum of the second order moment of mass distribution along the distance from the hair base. Therefore, we need to know the exact shape of the wind receptor hair, that is, the relationship between the diameter and the distance from the hair base. The geometrical configuration of the cercal hairs was examined by scanning electron microscopy (SEM).

The shaft diameters relative to the base size were plotted against the distance from the hair tip relative to the hair length, in a log-log diagram (Fig. 13.10B). The plotted points accumulated through all the measured hairs indicated a straight regression, which implies a power function between the two variables in abscissa and ordinate. If the regression slope is expressed as $1/m$, the relationship between the diameter d and the distance from the base y will be described as follows:

$d = d_0 \times \left(1 - \dfrac{y}{L_0}\right)^{\frac{1}{m}}$, or $m = \dfrac{\log(1 - y/L_0)}{\log(d/d_0)}$, where L_0 is the whole length of hair and d_0 is the base diameter.

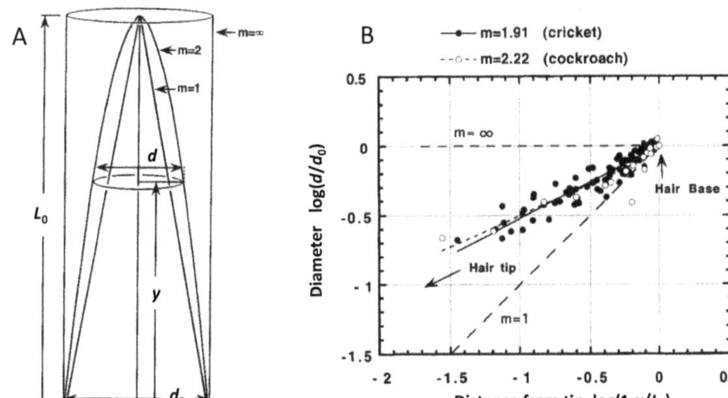

Figure 13.10 (A) Schematic diagram of the hair shape. A strait cylinder ($m = \infty$), a square root cone ($m = 2$), and a linearly tapered cone ($m = 1$) are drawn. Shapes are expressed by the shape index m, with which diameter d is given as a power function of height y, $d = d_0(1 - y/L_0)^{1/m}$. (B) Relationship between the diameter of the hair shaft and the distance from the hair tip. Filled dots and solid line: Crickets' cercal filiform hairs (117 data points). Open circles and dashed lines: Cercal filiform hairs of the cockroach, *Periplaneta americana* (22 data points). Lines with $m = \infty$ and $m = 1$ indicate the cases of strait cylinder and linearly tapered cone shapes.

The parameter m characterizes the shape of the wind receptor hair. The measurement of the geometrical configuration of the hair revealed that the hairs are very slender paraboloids or square root cones. The diameter of the hair shaft increases with the power of 1/2 of the distance from the hair tip, that is, $d(y) = d_0 \cdot (1 - y/L_0)^{0.52}$ [m] [9]. Since the allometric relationship between the base diameter d_0 and the hair length L_0 for hairs longer than 100 μm is known as $d_0 = 4.17 \times 10^{-4} \cdot L_0^{0.67}$ [m] (Fig. 13.2B), we can estimate both d_0 and $d(y)$ if the length L_0 of the hair subjected to mobility measurement is determined by microscope observation. By using these data, numerical integrations along the hair length for determining *A*, *B*, *C*, and *D* were obtained by Simpson's approximation formula. The most plausible set of three unknowns, *I*, *R*, and *S*, was determined by the least-squares error sum method applied to the 56–72 simultaneous equations at different frequencies and the $\pi/2$ phase difference for a single hair whose mobility was measured by LDV.

The geometrical estimation of the moment of inertia is now possible as,

$$I = \frac{m^3}{(m+1)(m+2)(3m+2)} d_0^{\ 2} \cdot L_0^{\ 3} \cdot \pi\rho_c, \ [\text{kg m}^2] \qquad (13.16)$$

where m (=1.92) is the shape index of the square root cone, L_0 is the hair length, d_0 is the base diameter ($= 4.17 \times 10^{-4} \cdot L_0^{0.67}$), and ρ_c is the density of the cuticle (=1.1×10^3 kg m^{-3}) [7, 20].

13.8 Allometric Scaling of Biological Design

The moment of inertia *I* obtained by the dynamic estimation showed a clear length dependency to the power of 4.32 of the hair length (Fig. 13.11). The dynamic estimation nicely fits with those obtained by geometrical estimation based on the hair shape of the squared root cone, for hair length in the range of 160–1500 μm [17]. As the moment of inertia of a cone-shaped hair is given by Eq. 13.16, the geometrical estimation should vary as a power function of hair length with an exponent of 4.34, very close to the observed value of 4.32. This agreement of the two independent estimations strongly supports the validity of the dynamic estimation of the mechanical parameters.

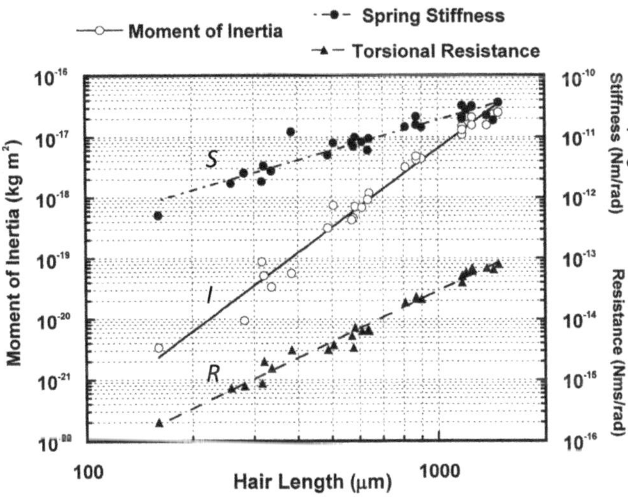

Figure 13.11 Three mechanical parameters of a cricket's wind receptor hairs plotted in relation to hair length show a clear allometric design. (*I*, moment of inertia of the hair shaft; *S*, stiffness of the hair supporting spring; and *R*, resistance within the hair base.)

The dynamic estimation also showed clear length dependencies of the stiffness of the hair supporting spring S and the torsional resistance R within the hair base, to the power of 1.67 and 2.77, respectively (Fig. 13.11). The slope of length dependency of the spring stiffness also fits with the previous static measurement of the exponent of 1.54 [16], although the absolute value of the dynamic estimation is two times larger than that obtained by the static method. The estimated values of torsional resistance within the hair base are so large that the hairs are strongly damped nonoscillatory second order systems (estimated damping factor ≈ 1.46). The large internal resistance is consistent with the overall sigmoid shape of the phase shift curve ($\pi/2$ per decade slope, Figs. 13.7-A2 and 13.7-B2).

The wind receptor hair consists of three cells, a trichogen cell, which secretes the hair shaft cuticle; a tormogen cell, which provides the hair supporting spring (joint membrane); and a sensory cell, which receives the stimulus energy. Although the evolutionary history of selection pressures that worked on the three cells is unknown, hairs with characteristic allometry (length dependency) on the three mechanical parameters compose an array covering a wide frequency range (Table 13.1) [17].

Table 13.1 Comparison of allometric scaling between the isometric physics and the observed or expected biological design of the cricket wind receptor hair

Quantity	Isometric physics	Biological design	Refs.
d_0: Base diameter	$\propto L$	$L^{0.67}$ (obs.[a])	[15]
S: Spring stiffness	$\propto L^2$	$L^{1.67}$ (obs.)	[16, 17]
I: Moment of inertia	$\propto L^5$ (=$d_0^2L^3$)	$L^{4.32}$ (obs.)	[17]
		$L^{4.34}$ (exp.[b] as $d_0^2L^3$)	[16, 17]
R: Resistance in hair base		$L^{2.77}$ (obs.)	
D: Air-hair resistance	$\propto L^3$ (theoretical)		[17]
ω_n: Natural frequency	$\propto L^{-1.5}$ (= $\sqrt{S/I}$)	$L^{-1.32}$ (exp. as $\sqrt{S/I}$)	[17]
f_b: Best frequency	$\propto L^{-1.5}$ (=0.38ω_n)	$L^{-0.92}$ (obs.)	[8]
θ_{max}: Best sensitivity	$\propto L^0$ (=A/S)	$L^{0.54}$ (obs.)	[8]
		$L^{0.33}$ (exp. as A/S)	[17]

[a]obs.: Observed
[b]exp.: Expected

13.9 Impedance Matching

The length dependency of the internal resistance R was estimated to the power of 2.77. This is fairly close to the theoretical expectation of a power of 3 for the length dependency of the frictional resistance D at the site of air-hair contact [17]. The coupling resistance D is given by an integral of the real part of Stokes's impedance for an infinitesimally thin cylinder along its length (Eq. 13.15). The large length dependence of D is a physical constraint for efficient energy transmission to the sensory cell. The power supply to R is maximum when R is the same as D, the condition called impedance matching.

The biological design seems to have attempted to match R (length to the power of 2.77) with D, although exact matching has not yet been achieved. The internal resistance seems to be in the impedance matching with frictional resistance at the site of air-hair contact, through which mechanical energy is transferred from air to hair. The energy transmission from the moving air to the sensory cell is maximized by the impedance matching.

The estimated length dependency of the best frequency is to the power of −0.92 and that of best sensitivity is 0.54 (Table 13.1). The cricket wind receptor hairs with different sizes compose an array of bandpass filters with different best frequencies. The wind receptor hairs are designed so as to shift the frequency band two decades while keeping a similar sensitivity. They exhibit a combination of allometric scaling of structural elements: the diameter of hair base, the spring stiffness, and the internal resistance.

The impedance matching, which provides the maximum energy transmission from moving air to sensory cell, seems to be a biological adaptation for higher energy sensitivity. Extension of this discussion leads to another physical understanding related to the resistance R, that R is not a simple frictional resistance but may represent absorption of the mechanical energy by the sensory cell for stimulus transduction. Sensation is a measurement that brings information about external world. Because information provides an ordered structure rather than random uncertainty, it is a type of negative entropy [13, 14, 18]. Although the second law of thermodynamics allows entropy increase, only a fraction of the mechanical energy absorbed by the sensory cell can be converted into information, if a much larger amount of biological energy stored within the living cell is released by the stimulus. Dissipation of the released biological energy compensates the entropy debt.

13.10 Energy Available to the Sensory Cell

The amount of energy consumed by the resistance R during one period of hair motion is given by $R \cdot (2\pi f \cdot \Theta)^2/(2f)$, where Θ and f are the deflection and frequency of hair motion, respectively. The peak amounts of kinetic and potential energy stored in and released from the moment of inertia and the spring stiffness during hair motion were similarly calculated as $(1/2) \cdot I \cdot (2\pi f \cdot \Theta)^2$ and $(1/2) \cdot S \cdot \Theta^2$, respectively. The moment of inertia of the air mass stagnating on the hair shaft due to viscosity was taken into account for calculation of the kinetic energy.

For the estimation of the amount of mechanical energy available to the sensory cell at sensory threshold, the deflection angle of the hair at sensory threshold is necessary in addition to the mechanical parameters obtained above. The deflection angle of hair at the sensory threshold (Θ_t) can be determined by combining two measurements, the sensory threshold in terms of wind velocity (Fig. 13.2A [15]) and the mobility of hair measured by LDV (Figs. 13.7-A1 and 13.7-B1). Deflections at the sensory threshold were frequency independent (Fig. 13.12) [18]. The sensory cells of a 1000 μm hair fire a spike when the hair is deflected by ca. 4×10^{-5} radians, regardless of the frequency of hair motion, whereas those of a 100 μm hair fire when deflected by ca. 4×10^{-4} radians.

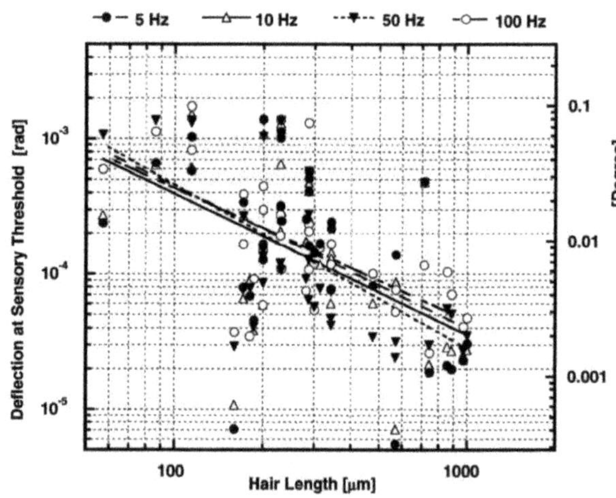

Figure 13.12 Hair deflection at the sensory threshold. The lines represent regressions at four different frequencies.

The amount of mechanical energy consumed by the torsional resistance within one cycle of hair motion was calculated for Θ_t (Fig. 13.13, filled dots). The amount of energy consumed by the resistance R provides the maximum estimate for the energy available to the sensory cell for activating a biological process of stimulus transduction. The energy threshold estimated is around k_BT of 300 K (4×10^{-21} joules, 25×10^{-3} eV), the energy level of the thermal fluctuation of a single molecule. This mechanoreceptor is therefore 100 times more sensitive than photoreceptors, which respond to a single photon quantum (ca. 3×10^{-19} joules). The cercal wind receptor of the cricket is the most sensitive cell ever reported. If any fraction of the torsional resistance measured is truly frictional, the amount of energy available to the sensory cell becomes much smaller.

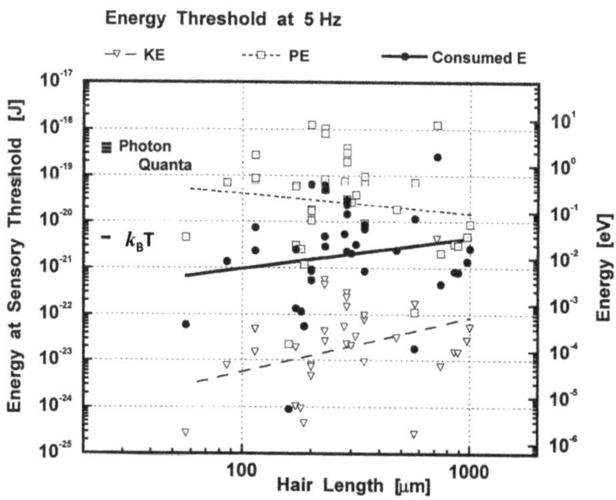

Figure 13.13 Maximum estimates of stimulus energy available to the sensory cell at the sensory threshold (filled dots), fairly close to thermal noise level k_BT at 300 K. (KE, kinetic energy stored in and released from the hair shaft mass motion; PE, potential energy stored in and released out from the hair support spring.

13.11 Conclusions, with Biological Insight into Engineering

The cricket's wind sensory cell has a mechanical sensitivity up to the ultimate physical limit of thermal noise, probably through the process of evolution. Although the sensory cells face thermal noise, they are still able to detect signals weaker than it. They utilize their intrinsic thermal noise for stochastic sampling as the source of noncorrelated noise, which contributes to signal detection. An ensemble of spiking sensory cells can transmit sensory information and reconstruct the stimulus waveform, if each member of the ensemble has an independent noise source [2, 3]. For a discussion about observation of a measurement system that contains thermal noise sources, see Refs. [13, 14]. Society demands electronic devices that are faster in operation speed, smaller in size, and lower in power consumption. To meet the demands, new devices such as single electron transistors and molecular or quantum devices are being developed. All these devices, however, are essentially constrained by noise or the stochastic nature of quantum mechanics. We must tame or cope with noise instead of trying to avoid it. Biological systems that have evolved under the inevitable thermal noise can teach us a design principle for future information systems that will face the same noise. Technology needs a wide and sound base in the natural sciences, not only physics, chemistry, and mathematics, but also biology.

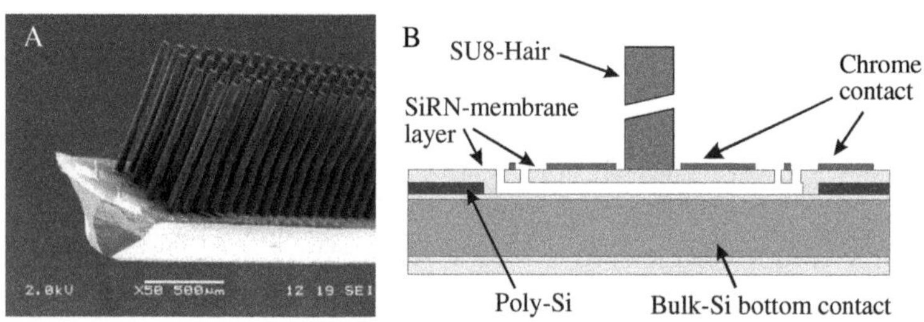

Figure 13.14 Artificial MEMS wind sensor: A hair shaft 1000 µm in length and 50 µm in diameter is made with superthick photoresist. The hair is supported by a spring of silicon nitride membrane. Hair deflection is detected by the change in the electrostatic capacitance between the chrome contact and the bottom base. Reprinted from Ref. [1] with permission of Jerome Casas.

At the end of the chapter, a man-made microelectromechanical systems (MEMS) wind sensor is introduced (Fig. 13.14). If a vector field of air velocity is observed by an sensor array, the moving object that caused the field of air flow is detectable without any reliance on vision [1, 4]. The most significant difference between the man-made, in-silico sensor and the cricket wind receptor is the mobility. The man-made sensor is hard, and mobility is small. Thus the amount of energy absorbed by the man-made sensor is very small compared to that of the biological soft interface with large mobility. The amount of information obtained by the physical measurement is, however, directly proportional to the amount of energy absorbed by the observer. More correctly, information is a negative entropy (absorbed energy divided by absolute temperature). Any amplification of measured signal cannot increase the amount of information at all. Future technology of the mechanical measurement device needs to match the impedance between sensor and subject of measurement.

Appendix

The Least-Squares Error Sum Determination of Mechanical Parameters

By using the measured values of the deflection amplitude Θ_m and φ_m of a hair at frequency f_m, we fixed the instantaneous values of angular acceleration, angular velocity, and angular deflection at time t_k as, respectively, $\alpha_{mk} = -(2\pi f_m)^2 \cdot \Theta_m \cdot \sin(2\pi f_m \cdot t_k + \varphi_m)$, $v_{mk} = (2\pi f_m) \cdot \Theta_m \cdot \cos(2\pi f_m \cdot t_k + \varphi_m)$, and $\theta_k = \Theta_m \cdot \sin(2\pi f_m \cdot t_k + \varphi_m)$. By replacing all time variables in the fundamental equation of hair motion with these values, we obtained a large number of equations for different times and for different frequencies after providing the values of torque N at each instance of time:

f_1: $$I \cdot \alpha_{11} + R \cdot v_{11} + S \cdot \theta_{11} - N_{11} = E_{11}$$

$$I \cdot \alpha_{12} + R \cdot v_{12} + S \cdot \theta_{12} - N_{12} = E_{12}$$

$$\vdots$$

$$I \cdot \alpha_{1k} + R \cdot v_{1k} + S \cdot \theta_{1k} - N_{1k} = E_{1k}$$

f_2: $$I \cdot \alpha_{21} + R \cdot v_{21} + S \cdot \theta_{21} - N_{21} = E_{21}$$

$$\vdots$$

f_m: $$I \cdot \alpha_{mk} + R \cdot v_{mk} + S \cdot \theta_{mk} - N_{mk} = E_{mk}$$

The total number of equations $n = m \times k$. Three unknown parameters, I, R, and S, are determined so as to minimize the sum of square error, $\sum_{i=1}^{n} E_i^2$. Thus,

$$\frac{\partial}{\partial I} \sum_{i=1}^{n} (I \cdot \alpha_i + R \cdot v_i + S \cdot \theta_i - N_i)^2 = 0,$$

$$\frac{\partial}{\partial R}\sum_{i=1}^{n}\left(I\cdot\alpha_i + R\cdot v_i + S\cdot\theta_i - N_i\right)^2 = 0,$$

and

$$\frac{\partial}{\partial S}\sum_{i=1}^{n}\left(I\cdot\alpha_i + R\cdot v_i + S\cdot\theta_i - N_i\right)^2 = 0.$$

Performing the partial differenntiations, we obtain a set of simultaneous algebraic equations for the three unknowns.

$$\left(\sum_{i=1}^{n}\alpha_i^2\right)\cdot I + \left(\sum_{i=1}^{n}v_i\cdot\alpha_i\right)\cdot R + \left(\sum_{i=1}^{n}\theta_i\cdot\alpha_i\right)\cdot S = \sum_{i=1}^{n}N_i\cdot\alpha_i$$

$$\left(\sum_{i=1}^{n}\alpha_i\cdot v_i\right)\cdot I + \left(\sum_{i=1}^{n}v_i^2\right)\cdot R + \left(\sum_{i=1}^{n}\theta_i\cdot v_i\right)\cdot S = \sum_{i=1}^{n}N_i\cdot v_i$$

$$\left(\sum_{i=1}^{n}\alpha_i\cdot\theta_i\right)\cdot I + \left(\sum_{i=1}^{n}v_i\cdot\theta_i\right)\cdot R + \left(\sum_{i=1}^{n}\theta_i^2\right)\cdot S = \sum_{i=1}^{n}N_i\cdot\theta_i$$

The solutions for *I*, *R*, and *S* are given by Cramer's formula for 3×3 matrix algebra.

References

1. Casas, J., Krijnen, G., Lammerink, T. and Wiegerink, R. (2008). *Insect Mimetics*, eds. Shimozawa, T. and Hariyama, T., Part 3, Chapter 1, Section 6; Cricket inspired MEMS flow-sensor arrays (NTS, Tokyo), pp. 657–668 (in Japanese).
2. Collins, J. J., Chow, C. C. and Inhoff, T. T. (1995). Stochastic resonance without tuning, *Nature*, **376**, pp. 236–238.
3. Denk, W. and Webb W. W. (1989). Thermal-noise-limited transduction observed in mechanosensory receptors of the inner ear, *Phys. Rev. Lett.*, **63**, pp. 207–210.
4. Droogendijk, H. (2014). Bio-inspired MEMS flow and inertial sensors, PhD Theses, University of Twente, Netherland.
5. Gnatzy, W. and Tautz, J. (1980). Ultrastructure and mechanical properties of an insect mechanoreceptor: stimulus transmitting structures and sensory apparatus of the cercal filiform hairs of Gryllus, *Cell Tissue Res.*, **213**, pp. 441–461.

6. Humphrey, J. A. C., Devarakonda, R., Iglesias, I. and Barth, F. G. (1993). Dynamics of arthropod filiform hairs. I. Mathematical modelling of the hair and air motion, *Philos. Trans. R. Soc. London, Ser. B*, **340**, pp. 423–444, with Errata added in *Philos. Trans. R. Soc. London, Ser. B* (1997) **352**, p. 1995.

7. Jensen, M. and Weiss-Fogh, T. (1962). Biology and physics of locust flight V. Strength and elasticity of locust cuticle, *Philos. Trans. R. Soc. London, Ser. B*, **245**, pp. 137–163.

8. Kumagai, T., Shimozawa, T. and Baba, Y. (1998). Mobilities of the cercal wind-receptor hairs of the cricket, *Gryllus bimaculatus*, *J. Comp. Physiol. A*, **183**, pp. 7–21.

9. Kumagai, T., Shimozawa, T. and Baba, Y. (1998). The shape of wind-receptor hairs of cricket and cockroach, *J. Comp. Physiol. A*, **183**, pp. 187–192.

10. Sakuranaga, M., Sato, S., Hida, E. and Naka, K.-I. (1986). Nonlinear analysis: mathematical theory and biological applications, *Crit. Rev. Biomed. Eng.*, **14**, pp. 127–184.

11. Shimozawa, T. (1991). Animal behavior and its neural basis, *J. Soc. Instrum. Contr. Eng.*, **30**, pp. 426–435 (in Japanese).

12. Shimozawa, T. (1993). Acoustic sense of insects, *J. Acoust. Soc. Jpn.*, **49**, pp. 413–420 (in Japanese).

13. Shimozawa, T. (2005–2006). Series Explanation: information theory for biologist - what is information? How is it measured? How does it explain the evolution of life?, *J. Jpn. Soc. Comp. Physiol. Biochem.*, **22**, pp. 32–37; **22**, pp. 85–89; **22**, pp. 149–154; **23**, pp. 32–36; **23**, pp. 38–43; **23**, pp. 154–164 (in Japanese).

14. Shimozawa, T. (2008). Why nerves are bundles of fibers?: information theoretic study on the basic structure of the nervous system, *Hyomen (Reviews & Topics on Surface Science and Technology Avant-Garde)*, **46**, pp. 551–564 (in Japanese).

15. Shimozawa, T. and Kanou, M. (1984). Varieties of filiform hairs: range fractionation by sensory afferents and cercal interneurons of cricket, *J. Comp. Physiol. A*, **155**, pp. 483–493.

16. Shimozawa, T. and Kanou, M. (1984). The aerodynamics and sensory physiology of range fractionation in the cercal filiform sensilla of the cricket *Gryllus bimaculatus*, *J. Comp. Physiol. A*, **155**, pp. 495–505.

17. Shimozawa, T., Kumagai, T. and Baba, Y. (1998). Structural scaling and functional design of the cercal wind-receptor hairs of cricket, *J. Comp. Physiol. A*, **183**, pp. 171–186.

18. Shimozawa, T., Murakami, J. and Kumagai, T. (2003). *Sensors and Sensing in Biology and Engineering*, eds. Barth, F. G., Humphrey, J. A. C. and Secomb, T. W., Chapter 10; Cricket wind receptors: thermal noise for the highest sensitivity known (Springer, Wien New York), pp. 145–157.

19. Stokes, G. G. (1851). On the effect of the internal friction of fluids on the motion of pendulums, *Trans. Camb. Phil. Soc.*, **9**, p. 8 (reprinted in *Mathematical and Physical Papers*, **III**, pp. 1–141, Cambridge University Press, 1901).

20. Tautz, J. (1977). Reception of medium vibration by thoracal hairs of caterpillars of *Barathra brassicae L.* (Lepidoptera, Noctuidae). I. Mechanical properties of the receptor hairs, *J. Comp. Physiol.*, **118**, pp. 13–31.

Chapter 14

Echolocation of Bats and Dolphins and Its Application

Ikuo Matsuo
Department of Information Science, Tohoku Gakuin University, Tenjinzawa 2-1-1, Izumi-ku, Sendai, 9813193, Japan
matsuo@mail.tohoku-gakuin.ac.jp

Bats and dolphins can use echolocation, which is called "biosonar." Bats emit high-frequency broadband sound waves to track and catch flying insects. They perceive the locations of these moving objects in 3D space. Also dolphins emit intermittently broadband ultrasound clicks. It is possible that such echolocation capabilities of bats and dolphins are applicable to the improvement of the artificial sounders that are used in acoustic surveys and so on. In this chapter, I describe the biomimetic model for a bat's echolocation and echo sounder system using the dolphin-like broadband signal.

14.1 Introduction

Bats and dolphins can use echolocation, which is called "biosonar." Bats emit high-frequency sound waves that enable them to track

Industrial Biomimetics
Edited by Akihiro Miyauchi and Masatsugu Shimomura

ISBN 978-981-4800-07-5 (Hardcover), 978-0-429-05883-7 (eBook)
www.jennystanford.com

and catch flying insects [1, 2]. They perceive the locations of these moving objects in 3D space using frequency modulation (FM). For example, Japanese house bats, *Pipistrellus abramus*, emit intermittently broadband ultrasounds, which sweep from about 100 kHz to 40 kHz within several milliseconds. It was shown that bats are capable of locating static objects at high signal-to-noise ratios (SNRs), achieving submicrosecond accuracy [3–6]. On the other hand, dolphins emit intermittently broadband ultrasound clicks of around 100 kHz with almost 100 μs duration. It was found that bottlenose dolphins (*Tursiops truncatus*) can determine the target size, material, and shape [7, 8] and object characteristics [9] from echoes by behavioral experiment. Both bats and dolphins can perceive objects by using broadband sonar signals. It is possible that such echolocation capabilities of bats and dolphins are applicable to the improvement of the artificial sounders that are used in acoustic surveys and so on. In Section 14.2, the biomimetic model of a bat's echolocation is described. In Section 14.3, the echo sounder system using the broadband signals is described while the conventional narrowband system has been used for the acoustic surveys in practical use.

14.2 Biomimetic Model of a Bat's Echolocation

Bats are capable of locating static objects with a range accuracy of less than 1 μs. The delay time provides information about the target's distance. In the case of range estimation using the cross correlation function between the emitted pulse and echoes, the range accuracy is also affected by the Doppler shift, which is dependent of the movement of objects and bats [10, 11]. Several models have been proposed that estimate the delay times of multiple objects from the echo spectrogram, which is computed using convolution of the constant frequency carrier wave [12–14]. However, it is difficult to accurately determine the distance for each object using the peak time because the integration time of the cochlear filters is large. Therefore, the time-frequency pattern is computed by convolution of the Gaussian chirplet filters for which the carrier frequency agrees with the sweep rate of emission [15, 16]. In general, the range accuracy of this was shown to depend on the SNR and frequency bandwidth of the emitted pulse [17–19]. It is necessary to estimate the distance of objects from measured echoes with a range accuracy of less than 1 μs at a low SNR.

14.2.1 Time-Frequency Analysis Mimicking the Auditory Peripheral System

The echo waveforms were entered into the model. The broadband FM emission was used to identify the location of the object. The 2-ms-long single FM signal started at 135 kHz and swept down to 5 kHz [15, 20]. The rise and fall times were both 50 μs. These echo waveforms were transformed into spectrograms in a manner that simulated the process in the mammalian cochlea. The Gaussian chirplet filters with a carrier frequency consistent with the sweep rate of emission have been used as this transform in order to extract the temporal changes in the interference pattern [15, 16]. The temporal characteristics of the filter can be described by the following equation:

$$F(f_j,t)=\exp\left(-\frac{t^2}{\alpha}\right)\exp\left(2\pi j\left(f_j t+\frac{1}{2}st^2\right)\right) \tag{14.1}$$

Here, f_j (kHz) is the center frequency for the jth bandpass filter, t is the time (s), s is the sweep rate of the emitted wave, and α is the parameter that describes the width of the window function. The bandpass filter bank comprised 71 filters with center frequencies ranging from 30 to 100 kHz, positioned at regular intervals. At 70% of the peak height, the width of the window function for all filters was 160 μs, with the quality factor at 10 dB ($Q_{10\text{dB}}$) values ranging from 1.5 at 30 kHz to 5.1 at 100 kHz. The waveforms for both the emitted waves and the echoes were transformed into the spectrogram by convolution with this filter, as shown in Eq. 14.1. The solid and dotted curves in Fig. 14.1 represent the spectrograms computed by the convolution of the Gaussian chirplet and the Gaussian tone filter corresponding to short Fourier transform, respectively. The temporal changes could be extracted by using these chirplet filters. The shapes of the temporal patterns corresponding to the spectrogram in all filters were the same due to use of the linear FM emission. The spectrogram, which was computed from the outputs of the Gaussian chirplets, was transformed into a range-frequency pattern with 10 μs intervals by compensating for the sweep rate of the emission FM. The compensation time was represented by τ, which implied that the range corresponding to the delay time since the start time of emission was 0.

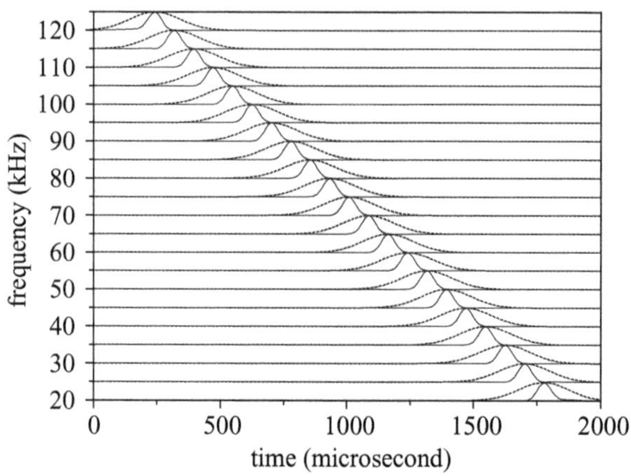

Figure 14.1 Emission spectrogram. The solid and dotted curves represent the spectrograms computed by the convolution of Gaussian chirplet and Gaussian tone filters, respectively. Reprinted with permission from Ref. [16]. Copyright 2004, Acoustic Society of America.

14.2.2 Model for Localization of Objects

The distance of one object was estimated from the range-frequency pattern around the onset [15, 20]. First, the averaged pattern was computed from the range-frequency pattern by the integration of bandpass filter channels. The delay times for the onset and offset were determined using a threshold [20–22]. The distance of the object was uniquely determined by extracting the temporal changes from the averages of the two spectra at the onset delay time τ_{on} and 10 μs later because this average pattern was dependent on the distance and reflectivity of the object [15].

The location of the object in 2D space was determined by the difference between the object's ranges at two microphones. The object's position (x, y) was represented by the polar coordinates (r, θ) by the following equations:

$$x = r\sin(\theta) \tag{14.2}$$

and

$$y = r\cos(\theta), \tag{14.3}$$

where r was the distance between the speaker and the object and θ the direction of the object along the horizontal axis. The distance r between the speaker and the object was computed by the mean between the ranges at two microphones. If $r >> d$, corresponding to the distance between two microphones, the direction of the object, θ, was computed from this difference, Δr, using the following approximate equation:

$$\theta = \sin^{-1}\left(\frac{\Delta r}{d}\right) \tag{14.4}$$

The object was continuously tracked by estimating its position at each timepoint.

14.2.3 Localization of the Static Object

To evaluate the range accuracy of this model, the echo from one object was measured. The distance (y) of one object from the loudspeaker was changed from 480 mm to 520 mm in steps of 1 mm in order to evaluate the range accuracy of the model. The closed circles and dotted line in Fig. 14.2a show the outputs of estimation and the actual distance, respectively, in the case of one object the position of which was changed along the y axis. Figure 14.2b shows the estimation errors, which were in agreement with the actual distance of the objects. It was clarified that the range of a single object could be estimated with high accuracy because the mean error was 0.45 mm, corresponding to 1.3 μs. In general, this range accuracy is dependent not only on the SNR but also on the frequency bandwidth [11, 18]. However, it was clarified that the accuracy was less than 2 mm, corresponding to about 6 μs, even when the bandwidths of the center frequencies were less than 10 kHz [20]. This model could therefore accurately estimate the range of one object using narrow frequency bandwidths by extracting the temporal changes.

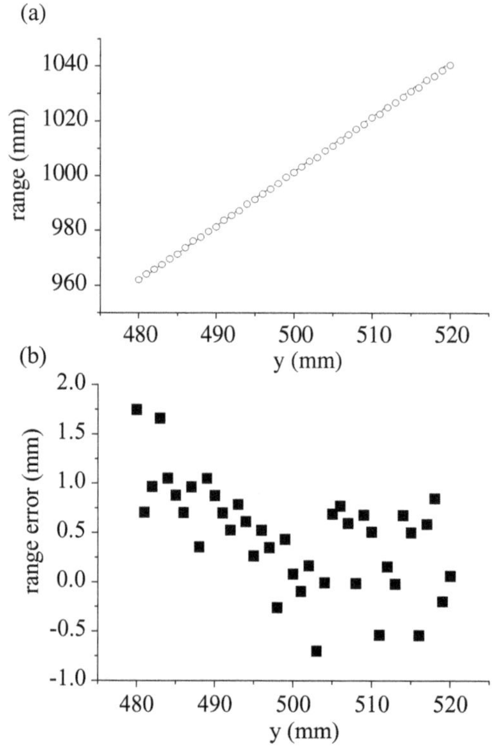

Figure 14.2 Estimation for one object. (a) Estimated range. (b) Estimation error. Reprinted with permission from Ref. [20]. Copyright 2011, Acoustic Society of America.

14.2.4 Localization of the Moving Object

The echoes from one object moving back and forth were measured and analyzed in order to evaluate the effect of the Doppler shift on the localization accuracy [21]. The center of rotation was fixed at (x,y) = (250 mm, 500 mm) and the radius of rotation was 250 mm. The circles in Fig. 14.3a show the estimated range along the time axis. In addition, the curves show the object's range and interaural range difference (IRD), which were computed by using the location of the moving object at each timepoint. The object's range could be estimated accurately using the temporal changes of echo spectra at the onset time. The circles and curves in Fig. 14.3b show the estimated location and position of the object in 2D space. It was clarified that locations of a moving object could be estimated using differences of the object's ranges for two microphones.

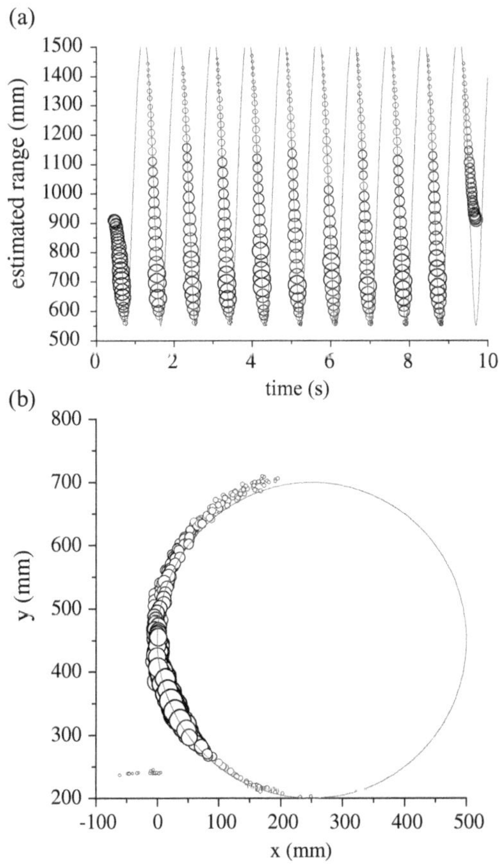

Figure 14.3 Estimation for one moving object. (a) Estimated and actual range. (b) Estimated and actual position in 2D space. Reprinted with permission from Ref. [21]. Copyright 2013, Acoustic Society of America.

14.3 Echo Sounder System Using a Dolphin-Like Sound

Acoustic surveys have been carried out for the estimation of fisheries resources. Narrowband ultrasound signals have been used for these surveys. Fish densities are typically estimated using measurements of echo amplitudes, that is, volume-backscattering strength of aggregations encountered directly below the survey vessel. Actually, fish density is estimated by the quotient of the total backscattered energy and the energy backscattered from an individual fish. However, it is difficult to estimate the density in a school of fish with several species of different sizes because the energy backscattered from an individual fish depends on its species and size. Therefore, it is necessary to improve the accuracy of the estimation. In contrast, dolphins make use of similar information when using broadband sonar signals to detect and pursue their prey. Behavioral experiments have found that bottlenose dolphins (*T. truncatus*) can determine the size, material, and shape of a target [7, 8], as well as the object characteristics [9]. It is therefore possible to apply the principles underlying the sonar capability of dolphins to improve the echo sounders that are used in acoustic surveys. The broadband split-beam echo sounder using a dolphin-like signal has been constructed to estimate fish abundance. The number of fish can be counted directly by using high range resolution to track individual fish by using this application.

14.3.1 Echoes from Fish

Figure 14.4 shows X-ray images in the lateral aspect of an anesthetized fish. It is known that the shape of the swim bladder varies depending on the fish species. The echoes from individual fish contain multiple reflections from the swim bladder and other organs [23, 24]. It is known that the temporal pattern of the echo from anesthetized fish is highly variable and depends on both the distance and the amplitude of these reflections [24–26]. These temporal and spectral patterns were shown to be correlated with the orientation of the fish [24, 27–28]. In addition, these patterns were found to be useful cues for discriminating fish species using the broadband signals emitted by dolphins [29–33].

14.3.2 Tracking Individual Fish Using the Split-Beam System

Dolphin sonar signals are composed of trains of clicks. Figure 14.5a shows the waveform of the click produced by a bottlenose dolphin (*T. truncatus*) for echolocation [26]. The sonar sound had its peak energy at a frequency of around 100 kHz, and the −3 dB bandwidth was 60 kHz (ranging from 55 to 110 kHz; Fig. 14.4b). A broadband split-beam echo sounder includes four receiving points in order to estimate the direction of each fish by using the time difference of arrival. The correlated signals were obtained by cross correlation between the emitted signal and the measured signals in each receiving channel. The position of a detected target was estimated from the range and the direction of arrival (DOA) of the echo [34, 35]. The alongship and athwartship angles were estimated by using the arrival time differences measured by the split-beam echo sounder. The half-beam signals were formed by summing two adjacent channels of the signals. The phase differences between the two signals received on the transducer half-beam pairs were then calculated via fast Fourier transforms. These phase differences in each frequency bins were then converted into time differences. A time difference was obtained using weighted average with the amplitude spectrum of the received signal. The DOAs of the echoes were estimated by using the time differences.

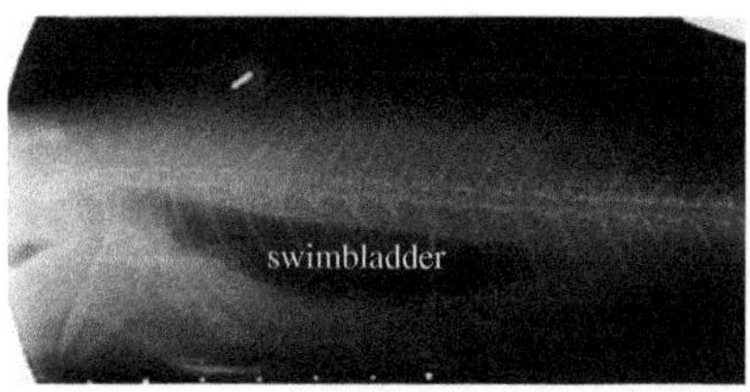

Figure 14.4 X-ray images in the lateral aspect.

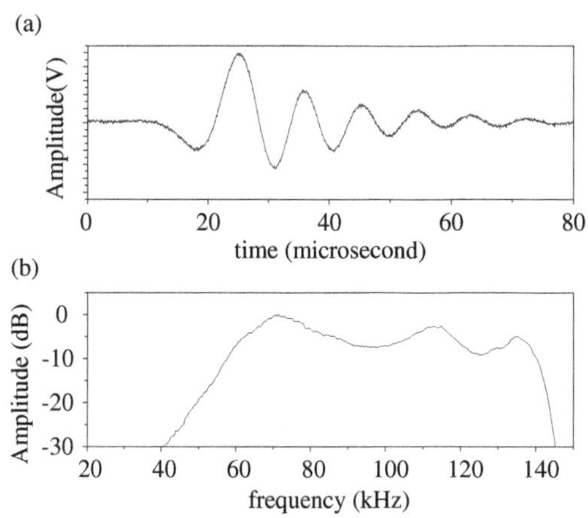

Figure 14.5 (a) Dolphin sonar sounds. (b) Amplitude spectra of dolphin sounds.

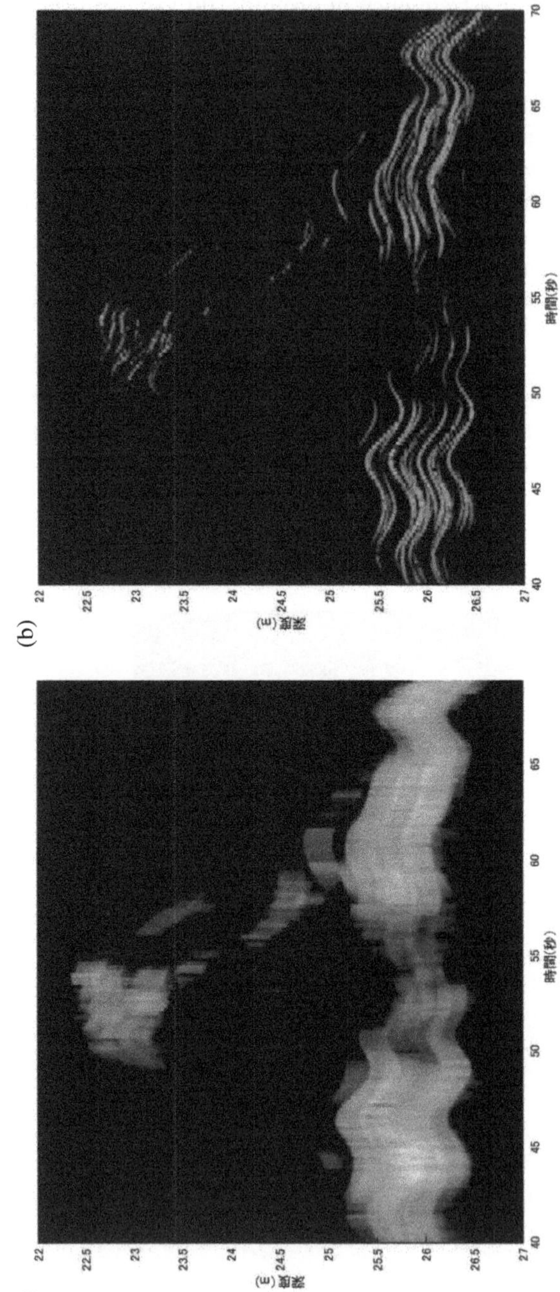

Figure 14.6 Echogram from a fish school. (a) Conventional narrowband system. (b) Broadband split-beam system.

14.3.3 Application in a Fish School

Figures 14.6a and 14.6b show examples of echograms from a school of Japanese anchovy measured by the conventional narrowband system and the broadband split-beam system, respectively [34]. The broadband system could isolate most of the echoes, whereas the conventional narrowband system produced overlapping echoes. Individual fish in the school were tracked as connecting the locations in 3D space along the time axis. Figures 14.7a and 14.7b show the tracking results in the top-down view and side view, respectively. Each line denotes movement of the individual fish in 3D space. The behavior of the fish could therefore be visually displayed.

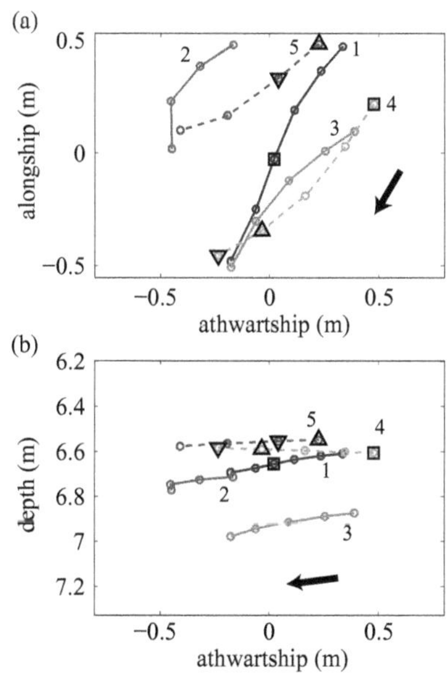

Figure 14.7 Fish movements estimated from localization and tracking. (a) Top-down view. (b) Side view.

14.4 Summary

Bats and dolphins can use echolocation, called "biosonar." They emit intermittently broadband sounds and perceive objects in 3D space. The biomimetic model for a bat's echolocation could localize the object by the time-frequency analysis, mimicking the bat's peripheral system, and extracting the temporal change of magnitudes. The range accuracy of this model was not dependent on the frequency bandwidth and was almost 0.45 mm, corresponding to 1.3 μs. Therefore, this model was useful for highly accurate range estimation. So, it is possible to apply the new biomimetic system in the future.

Also acoustic surveys have been carried out for the estimation of fisheries resources. Narrowband ultrasound signals have been used for these surveys. In this chapter, the broadband echo sounder system using a dolphin-like sound was described. This system has four receiving points, which is called the split-beam system. Individual fish could be accurately localized in 3D space by using the broadband signal. Individual fish in the school could be tracked by the dolphin-like sound.

Acknowledgments

Part of this work was supported by the Research and Development Program for New Bio-industry Initiatives of the Bio-oriented Technology Research Advancement Institution, Japan, and JST CREST Grant Number JPMJCR11A1, Japan.

References

1. Griffin, D. R., Friend, J. H. and Webster, F. A. (1965). Target discrimination by the echolocation of bats, *J. Exp. Zool.*, **158**, pp. 155–168.
2. Simmons, J. A., Ferragamo, M. J., Saillant, P. A., Haresign, T., Wotton, J. M., Dear, S. P. and Lee, D. N. (1995). Auditory dimensions of acoustic images in echolocation, in eds. Popper, A. N. and Fay, R. R., *Hearing by Bats* (Springer, New York), pp. 146–190.
3. Menne, D., Kaipf, I., Wagner, I., Ostwald, J. and Schnitzler, H. U. (1989). Range estimation by echolocation in the bat *Eptesicus fuscus*: trading of phase versus time cues, *J. Acoust. Soc. Am.*, **85**, pp. 2642–2650.
4. Moss, C. F. and Schnitzler, H. U. (1989). Accuracy of target ranging in echolocating bats: acoustic information processing, *J. Comp. Physiol. A*, **165**, pp. 383–393.
5. Simmons, J. A. (1979). Perception of echo phase information in bat sonar, *Science*, **207**, pp. 1336–1338.

6. Simmons, J. A., Ferragamo, M. J., Moss, C. F., Stevenson, S. B. and Altes, R. A. (1990). Discrimination of jittered sonar echoes by the echolocating bat, Eptesicus fuscus: the shape of target images in echolocation, *J. Comp. Physiol. A*, **167**, pp. 589–616.

7. Hammer, Jr. C. E. and Au, W. W. L. (1980). Porpoise echo-recognition: an analysis of controlling target characteristics, *J. Acoust. Soc. Am.*, **68**, pp. 1285–1293.

8. Au, W. W. L. (1993). *The Sonar of Dolphins* (Springer-Verlag, New York, NY).

9. Harley, H. E., Putman, E. A. and Roltblat, H. L. (2003). Bottlenose dolphins perceive object features through echolocation, *Nature*, **424**, pp. 667–669.

10. Altes, R. A. and Titlebaum, E. L. (1970). Bat signals as optimally doppler tolerant waveforms, *J. Acoust. Soc. Am.*, **48**, pp. 1014–1020.

11. Burdic, W. S. (1968). *Radar Signal Analysis* (Prentice-Hall, Englewood Cliffs, NJ), pp. 1–324.

12. Matsuo, I., Tani, J. and Yano, M. (2001). A model of echolocation of multiple targets in 3D space from a single emission, *J. Acoust. Soc. Am.*, **110**, pp. 607–624.

13. Neretti, N., Sanderson, M. I., Intrator, N. and Simmons, J. A. (2003). Time-frequency model for echo-delay resolution in wideband biosonar, *J. Acoust. Soc. Am.*, **113**, pp. 2137–2145.

14. Saillant, P. A., Simmons, J. A., Dear, S. P. and McMullen, T. A. (1993). A computational model of echo processing and acoustic imaging in frequency-modulated echolocating bats: the spectrogram correlation and transformation receiver, *J. Acoust. Soc. Am.*, **94**, pp. 2691–2712.

15. Matsuo, I., Kunugiyama, K. and Yano, M. (2004). An echolocation model for range discrimination of multiple closely spaced objects: transformation of spectrogram into the reflected intensity distribution, *J. Acoust. Soc. Am.*, **115**, pp. 920–928.

16. Matsuo, I. and Yano, M. (2004). An echolocation model for the restoration of an acoustic image from a single-emission echo, *J. Acoust. Soc. Am.*, **116**, pp. 3782–3788.

17. Boonman, A. and Ostwald, J. (2007). A modeling approach to explain pulse design in bats, *Biol. Cybern.*, **97**, pp. 159–172.

18. Menne, D. and Hackbarth, H. (1986). Accuracy of distance measurement in the bat Eptesicus fuscus: theoretical aspects and computer simulation, *J. Acoust. Soc. Am.*, **79**, pp. 386–397.

19. Simmons, J. A., Neretti, N., Intrator, N., Altes, R. A., Ferragamo, M. J. and Sanderson, M. I. (2004). Delay accuracy in bat sonar is related to the reciprocal of normalized echo bandwidth, or *Q*, *Proc. Natl. Acad. Sci. U. S. A.*, **101**, pp. 3638–3643.
20. Matsuo, I. (2011). Evaluation of the echolocation model for range estimation of multiple closely spaced objects, *J. Acoust. Soc. Am.*, **130**, pp. 1030–1037.
21. Matsuo, I. (2013). Localization and tracking of moving objects in two-dimensional space by echolocation, *J. Acoust. Soc. Am.*, **133**, pp. 1151–1157.
22. Matsuo, I. (2013). Echolocation of static and moving objects in two-dimensional space using bat-like frequency-modulation sound, Front. Physiol., **4**, p. 149.
23. Foote, K. G. (1980). Importance of the swimbladder in acoustic scattering by fish: a comparison of gadoid and mackerel target strength, *J. Acoust. Soc. Am.*, **67**, pp. 2084–2089.
24. Reeder, D. B., Jech, J. M. and Stanton, T. K. (2004). Broadband acoustic backscatter and high-resolution morphology of fish: measurement and modeling, *J. Acoust. Soc. Am.*, **116**, pp. 747–761.
25. Barr, R. (2001). A design study of an acoustic system suitable for differentiating between orange roughly and other New Zealand deep-water species, *J. Acoust. Soc. Am.*, **109**, pp. 164–178.
26. Matsuo, I., Imaizumi, T., Akamatsu T., Furusawa M., ans Nishimori, Y. (2009). Analysis of the temporal structure of fish echoes using the dolphin broadband sonar signal, *J. Acoust. Soc. Am.*, **126**, pp. 444–450.
27. Stanton, T. K., Reeder, D. B. and Jech, J. M. (2003). Inferring fish orientation from broadband acoustic echoes, *ICES J. Mar. Sci.*, **60**, pp. 524–531.
28. Au, W. W. L., Horne, J. K. and Jones, C. (2010). Basis of acoustic discrimination of chinook salmon from other salmons by echolocating Orcinus orca, *J. Acoust. Soc. Am.*, **128**, pp. 2225–2232.
29. Au, W. W. L. and Benoit-Bird, K. J. (2003). Acoustic backscattering by Hawaiian lutjanid snappers. II. Broadband temporal and spectral structure, *J. Acoust. Soc. Am.*, **114**, pp. 2767–2774.
30. Benoit-Bird, K. J., Au, W. W. L. and Kelley, C. D. (2003). Acoustic backscattering by Hawaiian lutjanid snappers. I. Target strength and swimbladder characteristics, *J. Acoust. Soc. Am.*, **114**, pp. 2757–2766.

31. Benoit-Bird, K. J., Au, W. W. L., Kelley, C. D. and Taylor, C. (2003). Acoustic backscattering by deepwater fish measured in situ from a manned submersible, *Deep Sea Res. Part I*, **50**, pp. 221–229.

32. Au, W. W. L. and Benoit-Bird, K. J. (2008). Broadband backscatter from individual Hawaiian mesopelagic boundary community with implications for spinner dolphin foraging, *J. Acoust. Soc. Am.*, **123**, pp. 2884–2894.

33. Yovel, Y. and Au, W. W. L. (2010). How can dolphins recognize fish according to their echoes? A statistical analysis of fish echoes, *PLoS One*, **5**, p. e14054.

34. Ito, M., Matsuo, I., Imaizumi T., Akamatsu, T., Wang, Y. and Nishimori, Y. (2014). Tracking individual fish in a dense school with a broadband split-beam system, *J. Mar. Acoust. Soc. Jpn.*, **41**(4), pp. 169–182.

35. Ito, M., Matsuo, I., Imaizumi, T., Akamatsu, T., Wang, Y. and Nishimori, Y. (2015). Target strength spectra of tracked individual fish in schools, *Fish. Sci.*, **81**, pp. 621–633.

Chapter 15

The "NanoSuit®" Preserves Wet/Living Organisms for Observation in High Resolution under a Scanning Electron Microscope

Takahiko Hariyama
Department of Biology, Hamamatsu University School of Medicine, 1-20-1, Handayama, Higashi-ku, Hamamatsu, Shizuoka 431-3192, Japan
hariyama@hama-med.ac.jp

Life and empty vacuums do not mix well. For wet/living organisms in particular, entering a vacuum means a ghastly death by dehydration. Electron microscopes can generate detailed images of the nanosized world by shooting electrons at their targets and detecting how the electrons react. However, because air molecules absorb the electrons and ruin the resolution, almost all scientists have to observe in vacuum. This ghastly vacuum is nearly as empty as outer space, and living organisms die quickly of dehydration as the vacuum sucks gases and liquids out of their bodies; therefore, all samples routinely require sacrifice and dehydration before they can be observed. Here

Industrial Biomimetics
Edited by Akihiro Miyauchi and Masatsugu Shimomura

ISBN 978-981-4800-07-5 (Hardcover), 978-0-429-05883-7 (eBook)
www.jennystanford.com

we describe the life-protection technique to overcome this limitation. With a nanothin layer, "NanoSuit®," organisms can be protected so they can be observed under an electron microscope.

15.1 Introduction

About 20 years after the invention of the scanning electron microscope (SEM) for nonorganic materials [1], biological observations began, and continuous growth and development continued for approximately 70 years. In 1965, the first commercial SEM was built by the Cambridge Instrument Company.

The observation technique using an SEM has become more versatile; therefore, the SEM is now an indispensable tool for advanced research across several scientific disciplines, including biology. Many paradigm shifts have occurred, especially the findings of viruses and organelles. However, it is necessary to evacuate the specimen chamber in order to prevent scattering by molecules in the air because of the beam of electrons an electron microscope uses to illuminate the specimen. All biological samples routinely require sacrifice and dehydration before observation, because approximately 80% of all living organisms is water, which rapidly evaporates under high vacuum, leading to disruption and collapse of the structure. To preserve and stabilize the biological structure for conventional SEM observations, complex treatment is required, including chemical fixation and careful drying. Furthermore, nonconductive materials cannot be imaged in a conventional SEM because they are poor emitters of secondary electrons due to the low atomic number of living substances, such as carbon. Therefore, samples usually require an ultrathin coating of electrically conducting materials [2]. Consequently, many time-consuming procedures are required to observe biological samples. In addition to the traditional chemical treatment, researchers have tried to modify SEM procedures to allow lower vacuum levels, as in low-vacuum scanning electron microscopy or by use of an environmental SEM (ESEM) [3–6]. All such methods, however, have resulted in inferior resolution and, thus, less information. Therefore, for a long time no living samples were observed by SEM except for a few reports [7, 8].

To observe the living and/or wet specimen in an SEM, we invented a new method to cover the surface of the specimen with a thin extra layer, termed the "NanoSuit®," which can prevent the escape of gases and liquid. The simple surface modification by electron beams or plasmas can equip the surface of some multicellular organisms with the NanoSuit® and hence can keep them alive under the high vacuum (10^{-3}–10^{-6} Pa) conditions of a field-emission scanning electron microscope (FE-SEM) [9–12]. In this article, we summarize the role of the NanoSuit® (various living specimens were protected by the NanoSuit®) during SEM observations.

15.2 Finding of the NanoSuit®

15.2.1 High-Vacuum-Tolerant Animals: Polymerization of NanoSuit® Based on Natural Substances

It has been known that under relatively low vacuum conditions (10 to 10^{-2} Pa) some organisms can be observed by SEM without any pretreatment [8, 9]; therefore, we tried to find the nature of tolerance of living organisms to high-vacuum conditions. We introduced numerous living organisms belonging to various taxa directly into the harsh environment of an SEM to see how long they survived under a high vacuum (10^{-3} to 10^{-6} Pa). All of their structural integrity was destroyed by rapid evaporation under high vacuum, including that of *Drosophila* larvae (Fig. 15.1).

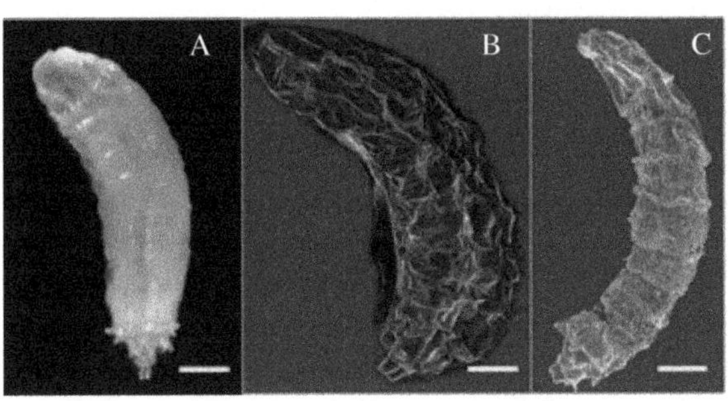

Figure 15.1 Light micrograph of a living larva (A), an SEM photo taken of a living larva introduced into the observation chamber of an SEM without electron beam irradiation for 30 min. (B), and an SEM photo of a traditional method. Scale bar 100 μm.

We, however, found that *Drosophila* larvae tolerated the high vacuum well when they were irradiated by an electron beam immediately in an SEM. Although they have a soft cuticle they continued to actively move around for 60 min. under the SEM (n = 25) (Fig. 15.2) and subsequently some of them (n = 18) developed normally.

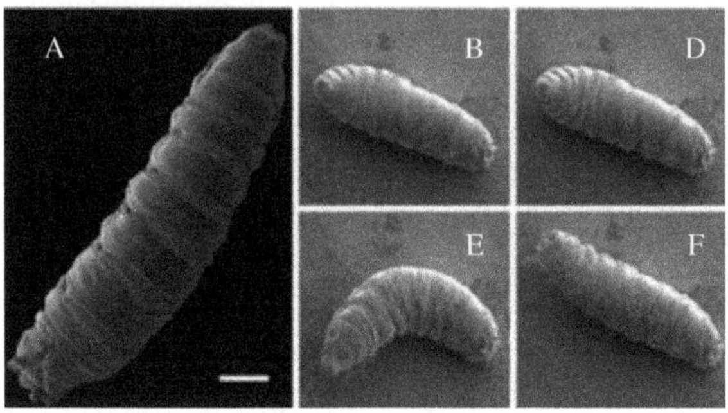

Figure 15.2 A living *Drosophila* larva irradiated by an electron beam immediately in high vacuum showed not only the live image (A) but also the active movement for 60 min. (B–F). Scale bar 100 μm.

To observe living organisms in an FE-SEM, it is necessary to provide a barrier to drastically reduce gas and liquid loss under high vacuum. We found that with the assistance of electron beam irradiation, a polymerized thin film NanoSuit® was produced and organisms acquired the potential to tolerate high-vacuum environments without any electric charges. This finding could shed light on new observations for various living specimens under a high-resolution SEM.

15.2.2 NanoSuit® Formation by Electron Beam and/or Plasma Irradiation

Why did only the larva of *Drosophila* survive after irradiation by an electron beam? Comparing with the animals, which died because of rapid evaporation under high vacuum, dipteran larvae possess a soft cuticle covered by extracellular substances (ECSs) [13]. The results described above led to the hypothesis that electron beam irradiation [14] enhances cross-linking within the ECSs to form a durable polymer on the surface and that this polymer increases resistance to vacuum conditions. To test this hypothesis, surface polymerization was achieved by a plasma irradiation. The ionized particles of plasma provide the energy necessary to initiate polymer formation, enabling uniform coating of surfaces with solvent-insoluble polymers [15, 16]. Although plasma polymerization so far has been used only in creating new inert industrial materials or modification of their surfaces [17], we applied this technique to the ECSs of living animals to construct a protective surface barrier referred to as the NanoSuit®. Plasma-irradiated specimens left in vacuo for 60 min. before SEM observation clearly showed features similar to the specimen electron beam irradiation by SEM ab initio: *Drosophila* larvae moved continuously, seemingly unharmed and without problems of electrical charging.

15.2.3 Mimicking ECSs for Other Animals

To mimic ECSs that could create the NanoSuit®, solutions including amphiphilic molecules were then tested. One of the best results obtained at the first stage was a solution of Tween 20, a nontoxic compound commonly used in biological experiments [18]. To test the barrier properties of the NanoSuit® made by this solution, the surfaces of certain animals previously unable to survive SEM exposure were provided exogenous materials by immersion in 1% Tween 20 solution before electron or plasma irradiation. Figure 15.3A–C shows typical results obtained with larvae of the Asian tiger mosquito, *Aedes albopictus*.

The larvae of this mosquito are aquatic and possess a soft-bodied cuticle with no apparent ECS. When live larvae of *A. albopictus* were observed under the SEM without any additional treatment, they quickly shrank and ceased to move. Larvae treated with a 1% Tween 20 solution but not irradiated in the SEM showed the same collapsed structure when observed 30 min. later. However, living larvae (Fig. 15.3A) covered with 1% Tween 20 and observed under an SEM ab initio retained their morphology and exhibited active movements for 30 min. (Fig. 15.3B). After the latter were reared for three to four days, approximately 80% pupated and developed successfully into adult mosquitoes (n = 25) (Fig. 15.3C).

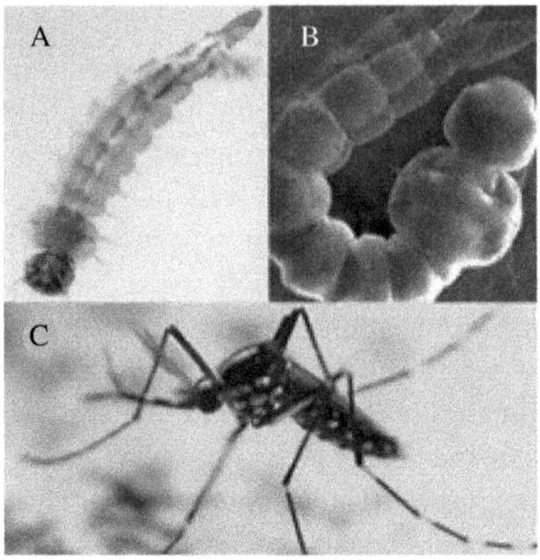

Figure 15.3 A living *A. albopictus* larva (A) was exposed to electron bombardment after quick immersion using Tween 20 solution. It moves in high vacuum for 60 min. The blur of the photo observed at the tail was caused by its fast movement (B). 80% of larvae observed under an SEM developed into adult mosquitoes (C).

15.2.4 Transparency of the NanoSuit® for Observations

To investigate how the NanoSuit® allows visualization of fine structures with good resolution under the SEM, we prepared NanoSuits® of varying thicknesses and compared the SEM images with transmission electron microscope (TEM) images of cross sections of the NanoSuit® using *Culex pipiens molestus* larvae treated with Tween 20. When the surface of the animal was covered by the NanoSuit®, the fine structure was visible even though the thickness of the NanoSuit® was about 100 nm and sometimes it fully filled the ridges. However, when the surface of the NanoSuit® was covered with gold by spattering, the fine structure of the ridges disappeared, which must be the reason the electrons cannot penetrate the thin layer of gold [11].

15.3 The Biomimetic ECS Films Made by Other Chemical Compounds

Covering the living organisms with a layer of the mimicked ECS, 1% Tween 20, polymerized by plasma or electron beam irradiation made the organisms highly tolerant to high vacuum. Can other biocompatible chemical compounds fabricate the NanoSuit® for living organisms? We investigated biomimetic ECS films made by plasma polymerization from surfactants, water-soluble polymers, monosaccharides, polysaccharides, lipids, amino acids, and ionic liquids [19]. Briefly, the monomer layer was spin-coated on a glass substrate and then placed in an irradiation chamber. The water-soluble layer, which consisted of the monomer and water, remained wet even under vacuum (0.13 Pa). Thin films were formed on the outermost surface of the monomer layer after plasma irradiation. The unreacted monomer was removed by immersing the plasma-irradiated films in a large volume of pure water and ethanol. The insoluble plasma-irradiated films were transferred to a substrate and allowed to dry under ambient conditions, demonstrating they were freestanding. The thickness of the insoluble film was controlled by adjusting several points: the electrical potential, the irradiation time, the distance to the irradiated surface, the concentration of the solution, and the spin-coating conditions.

15.4 Tissues and Cells with NanoSuits®

Although the Tween 20 based NanoSuit® has a surface shield effect and protects some organisms in an SEM, it was unable to protect isolated tissues and cells excised from intact organisms. To overcome this limitation, we have modified the technique and developed a new solution; a surface shield enhancer (SSE) solution that enables FE-SEM observations on wet samples.

15.4.1 Tissues

To explore the effect of the SSE-based NanoSuit® for "wet-fixed specimens," observations were carried out on chemically fixed pathology samples. Figure 15.4 shows images of a surgical explant of the human stomach wall including areas of normal and cancerous tissue. Morphological features were compared using two different preparations: specimens treated with the SSE-NanoSuit® solution and specimens prepared by conventional methods for SEM observation (fixation, freeze-drying, and metal coating). Specimens treated with the SSE-NanoSuit® solution (Fig. 15.4) appeared intact. By comparison, fixed specimens prepared by conventional metal coating for SEM showed obvious structural damage (data not shown).

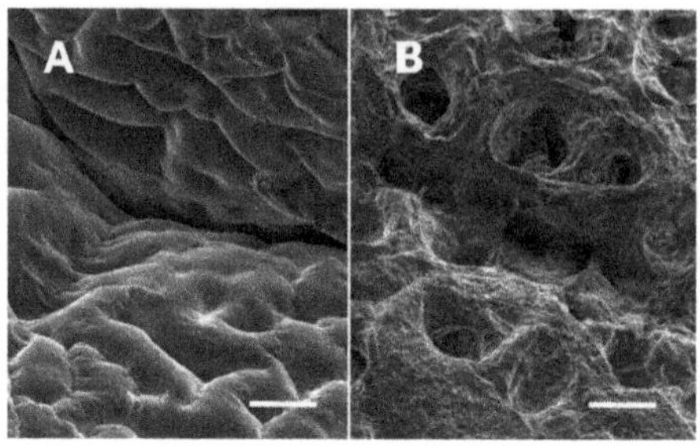

Figure 15.4 Comparison of normal (A) and cancerous stomach tissue (B) protected by an SSE-based NanoSuit®. Scale bar 100 μm.

15.4.2 Cells

To protect the integrity of single cells under high vacuum, numerous combinations of solutions were tested [19]. The best results were obtained with a new SSE-NanoSuit® solution—a mixture of glycerin and electrolytes. To test the barrier properties of the NanoSuit® based on this solution, the slide glasses with cultured cells were dipped in the SSE-NanoSuit® solution and then exposed to an electron beam or plasma irradiation. When these specimens were subsequently introduced into the FE-SEM they showed no observable change in appearance (Fig. 15.5D), suggesting that the SSE-NanoSuit® prevented desiccation of the samples. To quantify the ability of the SSE-based NanoSuit® to act as a barrier against gas and/or liquid loss, the weight loss of several samples was measured after 30 min. of observation under the high vacuum of the SEM. The untreated specimens and specimens treated with the Tween 20 solution lost roughly 50% of their weight after 30 min. whereas specimens treated with the SSE-NanoSuit® solution showed only 9% weight loss. Thus, the NanoSuit® formed with the SSE-NanoSuit® solution is an effective barrier against water loss.

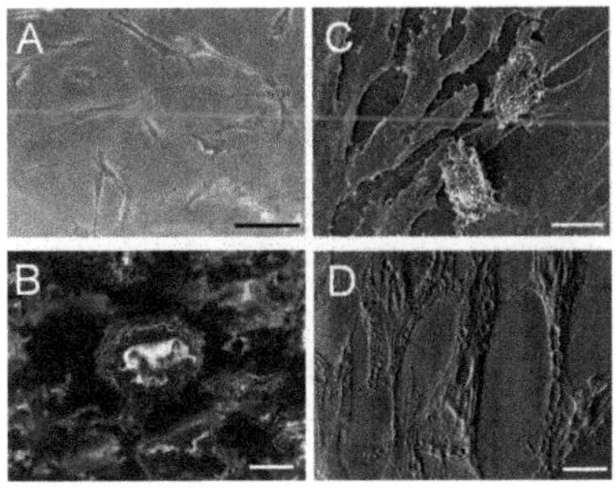

Figure 15.5 Comparison of images of a human fibroblast under light microscope (A), direct observation without any treatment (B), prepared by traditional methods (C), and treated with the SSE-NanoSuit® solution. Fibers and nucleus of the cells are apparent.

15.5 Discussion

To maintain life under natural conditions, water, which possesses many unique properties and plays an irreplaceable role in organisms, is essential. Thus, water limitation is one of the harshest stressors in an extreme environment, viz. a high vacuum. To observe living organisms using a conventional SEM in a high vacuum, there are two major problems to solve: how to produce a sufficiently thin barrier of gas and/or liquid through which to observe the surface fine structure and how to increase the electric conductivity of the surfacc of the animal without any toxicity or damage to the living tissues. We found that after modification of chemical compounds on the surface of organisms through exposure to an electron beam or plasma ionization, that is, conditions known to enhance polymer formation, some treated animals, such as the larvae of *Drosophila*, which possess ECSs, survived under the high-vacuum conditions of a conventional SEM, including an FE-SEM, due to the protective properties of what we termed the "NanoSuit®." From this result, we realized that, with the assistance of a polymerized thin film made of natural ECSs, organisms get the potential to tolerate high vacuum environments without any electrical charges. The advancement that we have made is to preserve life in high vacuum long enough to observe active movements of living specimens at a good resolution. In addition to this, organisms that lack ECSs were successfully protected with a Tween 20–based NanoSuit®. The results revealed that the natural ECSs and the artificial compounds play the same role in composing the NanoSuit® to protect the life of the whole organism in a harsh environment.

By contrast, excised tissues and cells do not have such a protective cover and the Tween 20 solution is harmful to naked cell membranes, so an alternative barrier was needed. We used an SSE-NanoSuit® solution with glycerin as a main component. Since glycerin is strongly hygroscopic, it has been used as a humectant in cosmetics. To increase the barrier effect, we have combined glycerin with electrolytes and polymerized a thin liquid film over the sample in order to prepare the SSE-NanoSuit® for naked tissues. Our results suggest that to protect naked tissues for long periods under high-vacuum conditions, this is a very effective barrier against the harsh environment.

The structures observed by the NanoSuit® method showed apparent differences from those obtained by the conventional method. This ongoing progress in electron microscopy is starting to have a significant impact on our understanding of the subcellular world. The NanoSuit® method will allow more sophisticated observation, revealing many life science worlds, from the subcellular materials to the whole organism, in the living state.

Acknowledgments

This work was supported by Grant-in-Aid for Scientific Research on Innovative Areas in "Innovative Materials Engineering Based on Biological Diversity" (JP24120001 and JP24120004), by JSPS KAKENHI (JP18H01869), and by grants from the Takeda Science Foundation. I wish to express my cordial thanks to all colleagues of the NanoSuit Research Group.

References

1. Knoll, M. (1935). Aufladepotential und Sekundäremission lektronenbestrahlter Körper, Z. Tech. Phys., 16, pp. 467–475.
2. Suzuki, E. (2002). High-resolution scanning electron microscopy of immunogold-labelled cells by the use of this plasma coating of osmium, J. Microsc., 208(Pt 3), pp. 153–157.

3. Danilatos, G. D. (1991). Review and outline of environmental SEM at present, J. Microsc., 162, pp. 391–402.

4. Mohan, A., Khanna, N., Hwu, J. and Joy, D. C. (1998). Secondary electron imaging in the variable pressure scanning electron microscope, J. Scan Microsc., 20(6), pp. 436–441.

5. Symondson, W. O. C. and Williams, I. B. (2003). Low-vacuum electron microscopy of carabid chemoreceptors: a new tool for the identification of live and valuable museum specimens, Entomol. Exp. Appl., 85(1), pp. 75–82.

6. Stokes, D. J. (2003). Investigating biological ultrastructure using environmental scanning electron microscopy (ESEM), in Science, Technology and Education of Microscopy: An Overview, pp. 564–570.

7. Pease, R. F., Hayes, T. L., Camp, A. S. and Amer, N. M. (1966). Electron microscopy of living insects, Science, 154, pp. 1185–1186.

8. Sokoloff, A., Hayes, T. L., Pease, R. F. W. and Ackermann, M. (1967). Tribolium castaneum: morphology of "aureate" revealed by the scanning electron microscope, Science, 157(3787), pp. 443–445.

9. Takaku, Y., Suzuki, H., Ohta, I., Ishi, D., Muranaka, Y., Shimomura, M. and Hariyama, T. (2013). A thin polymer, nano-suit, enhancing survival across the continuum between air and high vacuum, Proc. Natl. Acad. Sci. U. S. A., 110(19), pp. 7631–7635.

10. Ohta, I., Takaku, Y., Suzuki, H., Ishii, D., Muranaka, Y., Shimomura, M. and Hariyama, T. (2014). Dressing living organisms in a thin polymer membrane, a NanoSuit, for FE-SEM observation, Microscopy, 63(4), doi: 10. 1093/jmicro/dfu015.

11. Takaku, Y., Suzuki, H., Ohta, I., Tsutsi, T., Matsumoto, H., Shimomura, M. and Hariyama, T. (2014). A 'NanoSuit' surface shield successfully protects organisms in high vacuum: observations on living organisms in an FE-SEM, Proc. R. Soc. London, Ser. B, 282, p. 20142857.

12. Takaku, Y., Suzuki H., Kawasaki H., Ohta I., Ishii D., Hirakawa S., Tsutsui T., Matsumoto, H., Takehara, S., Nakane, C., Sakaida, K., Suzuki, C., Muranaka, Y., Kikuchi, H., Konno, H., Shimomura, M. and Hariyama, T. (2017). A modified 'NanoSuit®' preserves wet samples in high vacuum: direct observations on cells and tissues in field-emission scanning electron microscopy, R. Soc. Open Sci., 4, p. 160887.

13. Chambers, L., Woodrow, S., Brown, A. P., Harris, P. D., Phillips, D., Hall, M., Church, J. C. and Prichard, D. I. (2003). Degradation of extracellular matrix components by defined proteinases from the greenbottle larva Lucilia sericata used for the clinical debridement of non-healing wounds, Br. J. Dermatol., 148(1), pp. 14–23.

14. Yasuda, H. (1981). Glow discharge polymerization, J. Membr. Sci., 16(1), pp. 199–293.
15. Sun, K. H. (1954). Effects of atomic radiation on high polymers, Mod. Plast., 32(1), pp. 141–238.
16. Friedrich, J. (2011). Mechanisms of plasma polymerization – reviewed from a chemical point of view, Plasma Processes Polym., 8, pp. 783–802.
17. Jiang, C., Markutsya, S. and Tsukruk, V. V. (2004). Compliant, robust, and truly nanoscale free-standing multilayer films fabricated using a pin-assisted layer-by-layer assembly, Adv. Mater., 16, pp. 157–161.
18. FAO (1974). Toxicological evaluation of some food additives including anticaking agents, antimicrobials, antioxidants, emulsifiers and thickening agents, FAO Nutr. Meet. Rep. Ser., 53A, pp. 1–520.
19. Suzuki, H., Takaku, Y., Ohta, I., Ishii, D., Muranaka, Y., Shimomura, M. and Hariyama, T. (2013). In situ preparation of biomimetic thin films and their surface-shielding effect for organisms in high vacuum, PLoS One, 8(11), p. e78563.

Index

For Product Safety Concerns and Information please contact our EU representative GPSR@taylorandfrancis.com
Taylor & Francis Verlag GmbH, Kaufingerstraße 24, 80331 München, Germany

www.ingramcontent.com/pod-product-compliance
Ingram Content Group UK Ltd.
Pitfield, Milton Keynes, MK11 3LW, UK
UKHW022159190926
475702UK00003B/103

* 9 7 8 9 8 1 4 8 0 0 0 7 5 *